Tertiary Education
in the 21st Century

Tertiary Education in the 21st Century

Economic Change and Social Networks

Robert Craig Strathdee

palgrave
macmillan

First published in 2008 by
PALGRAVE MACMILLAN™
175 Fifth Avenue, New York, N.Y. 10010 and
Houndmills, Basingstoke, Hampshire, England RG21 6XS.
Companies and representatives throughout the world.

PALGRAVE MACMILLAN is the global academic imprint of the Palgrave
Macmillan division of St. Martin's Press, LLC and of Palgrave Macmillan Ltd.
Macmillan® is a registered trademark in the United States, United Kingdom
and other countries. Palgrave is a registered trademark in the European Union
and other countries.

ISBN-13: 978-1-4039-7517-1
ISBN-10: 1-4039-7517-5

Library of Congress Cataloging-in-Publication Data is available from the
Library of Congress.

A catalogue record of the book is available from the British Library.

Design by Scribe Inc.

First edition: August 2008

10 9 8 7 6 5 4 3 2 1

Printed in the United States of America.

Contents

Acknowledgments

The central aim of this book is to better understand the competition for advancement through tertiary education. I look critically at the role played by the reputations of providers of tertiary education in facilitating transitions into the labor market. I believe that if we are to advance our understanding of the issues, we need to look within relatively discrete fields.

The initial stimulation for this book came from earlier work published in the *Journal of Educational Policy* (Strathdee 2005). I developed the argument that it is the superior ability of elite universities to create innovative knowledge that helps explain why their graduates earn more than those who attend less prestigious institutions. The book tests this idea by interviewing managers in contrasting fields.

There are many people to whom I am indebted. In particular, I thank Amanda Johnson and Brigitte Shull, the commissioning editors from Palgrave Macmillan, for their support of the book. Thanks also are due to those who helped improve the manuscript through providing editorial assistance and critique. These include Jasmine Fremantle, Charlotte Clements, Madeleine Clements, and Susan Kaiser. The empirical chapters of the book are based on interviews that were conducted with managers in a variety of fields. Without their help, this book would not have been possible.

CHAPTER 1

Introduction

As the adage goes, it is not what you know but who you know that counts in finding employment. Far from being folk wisdom, the relationship between who you know and employment has been validated in numerous studies that have been conducted in diverse settings and have employed a variety of research methods (Granovetter 1995). For such reasons, networking is now advanced by government agencies, career advisors, and many other groups and individuals as an important strategy for finding employment. The importance of networking during processes of allocating labor power has been aided by developments in information technology that enable workers to connect to diverse and widespread networks (Castells 1998). It is also widely agreed that networks have an important role to play in the creation and dissemination of new knowledge throughout the economy. Thus, most conceptions of the knowledge economy emphasize the emergence of networks, which facilitate labor mobility, knowledge creation, and knowledge transfer.

For social and economic reasons, governments in England, New Zealand, and other Western nations have an interest in social networks. One reason for this interest is that these governments want to maximize the economic and social returns from their investment in knowledge creation. For example, policies such as the Research Assessment Exercise in England and the Performance-Based Research Fund in New Zealand are attempts to improve the economic returns from investing in knowledge creation by concentrating research and development capacity in relatively few institutions.[1] Spreading investments more widely would increase the likelihood of the duplicating effort and would mean that the state would struggle to fund key "blue sky" research. However, this book argues that governments in these nations are attempting to disperse the knowledge created by this research throughout the economy by social network formation in an attempt to reduce the likelihood that such policies contribute to the establishment of new forms of social closure.

The adage noted above presents who you know and what you know as conflicting premises. However, in practice, what and who are factors that interact with each other. As the work of social network theorists shows, social networks provide an important means of gaining knowledge that is embedded within the social infrastructure. Such knowledge can take many different forms, including tacit knowledge and codified knowledge (Nonaka and Konno 1998). Erickson (1996) argues convincingly that social networks are a vital source of knowledge and cultural resources that contribute to competitive advantage. According to Erikson, those with more diverse networks are best placed to gain the positional knowledge and develop the cultural capacity needed to win the advantage. Thus, access to embedded knowledge increases competitive advantage and contributes to processes of social class reproduction. For example, for organizations, access to embedded knowledge can provide information about new innovations and reduce the chances of making hiring errors by providing trustworthy and reliable information about the quality of potential recruits. For individuals, access to embedded knowledge improves the value of their human capital, and it also can increase their chances in the recruitment process by according them insider status.

Although knowing who and knowing what are clearly related, it is important to establish that the precise mix of these factors, and other assets needed for advancement, varies from setting to setting. In areas of the labor market in which formal qualifications play a limited role in the recruitment process, knowing who is likely to be more important than knowing what in terms of gaining employment. In such circumstances, employers are likely to rely more heavily on knowledge of who (and other signals of competency, such as the reputation of the tertiary education institution applicants attended) to recruit new staff. In other areas, such as those in which the human capital needed by employers can only be taught in institutions of tertiary education and possession of the required skills and qualities is effectively signaled by educational qualifications, knowing who is likely to be of reduced significance in the recruitment process.

A central theme of this book is that economic change is driving up the importance of network capital. For individuals, governments, and providers of education, these changes are significant because they are leading to a reconfiguration of the rules on which the competition for advancement is organized through tertiary education. The book aims to contribute to our understanding of some of the factors influencing network formation and to describe how the resulting changes in the provision of tertiary education are helping shape new rules of advancement (or, as shown later in the book, creating new geographies of talent). Achieving the book's central purpose requires a focus on how educational reform is forcing changes in the role played by tertiary education in both mediating and structuring the competition for advancement. This also requires a focus on the strategies used by managers to solve their human resource requirements. Although the book is primarily conceptual in nature, qualitative evidence is used to test the ideas advanced.

The starting point here is the argument that network formation is a key source of innovation and hence, competitive advantage. Of central importance is the way in which policy changes are encouraging knowledge generators, and particularly (but not exclusively) universities, to form informal and formal alliances with innovative firms and with sources of finance. In this respect, a key development in the contemporary period is the move toward an interactive model of innovation. In this model, knowledge flows and relationships between industry, government, and providers of tertiary education are encouraged by state policy (Archibugi and Lundvall 2001). These alliances have developed because firms, the state, providers of tertiary education, and individuals all can benefit from them. For example, firms can benefit through the improved competitiveness that is derived from accessing the research expertise within institutions of tertiary education and from sources of information about potential recruits that flow from network relationships. Similarly, tertiary institutions can benefit from the financial support provided by firms and from the employment of their graduates. A critical consideration here is the extent to which providers of tertiary education are able to transfer innovative knowledge (and other sources of competitive advantage) through their education and training programs. To maximize their contribution to competitive advantage, providers of tertiary education have to ensure that their students have the competencies and the knowledge needed to enable the firms in which they work to innovate.

As suggested earlier, the relationship between networks and competitive advantage only applies in particular areas of the labor market, and an exploration of the relationship between knowing who and knowing what requires an investigation of processes occurring within relatively discrete areas of the labor market. To facilitate this kind of exploration, the book adopts Bourdieu and Wacquant's notions of field, capital, and habitus (Bourdieu and Wacquant 1992). This approach might situate this book too completely within culturalist accounts of social class—and too far from economic accounts—for some (Scott 1996). Nevertheless, their approach does offer a way to understand how processes of social class reproduction vary throughout the labor market. In his model, the way in which class and culture interact depends on the field, or the social setting, in which the interaction occurs. For example, universities have been conceptualized as a field in which institutionalized forms of cultural capital, or academic capital, have currency (Bourdieu and Wacquant 1992).

Although conceiving of the competition for advancement as occurring in different ways in different settings offers an opportunity to improve our understanding, limitations remain. For example, even within the field of higher education, the capitals needed to advance within the field can vary greatly. As the debates that took place following Snow's (1964) two cultures lectures demonstrate, the knowledge structures found in the sciences differ from those found in the humanities, raising the possibility that the kinds of capital needed for advancement within the field of higher education differ from discipline to discipline. According to Snow, the ideal humanist was a gentleman amateur, and access to the humanities' scientific culture was limited to those who held the requisite habitus. In contrast, anyone could become a scientist, so long as they acquired technical expertise. Because expertise could be gained through formal education and training, the sciences were seen to be egalitarian, whereas the humanities were seen to be class-bound.

As described in more detail in the following chapters, adopting a field approach provides a way to address weaknesses in existing research, which tends to draw conclusions about the relationship between qualifications and the labor market across fields. A related weakness is that the research does not consider how the relationship between qualifications and the labor market might vary across time and across space (Gough 2004). Indeed, variability in the mix of resources needed for advancement highlights the importance of studying processes of social reproduction within particular social fields and alerts us to the fact that the rules of advancement operating within these are likely to change over time and space. For example, domestic networks might be of importance in some fields, but in others, international networks may be of importance. In terms of the precise configuration of networks, what matters are the competitive forces at play and the source of the innovations that give rise to competitiveness.

Although it clearly leads to competitive advantage in many fields, some people have distaste for networking, arguing convincingly that it is exclusionary and that it undermines meritocracy. It can be argued that networking reduces economic efficiency by advancing those with the right contacts, rather than those who are best suited for a given position. Attempts to promote equality and efficiency have led egalitarian reformers to exert much effort toward creating recruitment procedures and related legislation that create open competitions for advancement. For example, in education, the desire to create open competitions underpins the development of new forms of curriculum and assessment. In part, these are designed to create better forms of knowing what, or human capital, and to establish a more accurate and egalitarian way of measuring capacity, so that knowing who loses importance. In other words, such reforms are designed to create new, open geographies of talent.

However, despite the best efforts of egalitarians to erase the influence of knowing who on advancement, developments in the economy, education, and labor market are working against this aim by increasing the significance of network formation. Some innovative firms are organizing themselves around different principles than those that were dominant in the past. For example, rather than having total and direct control over the production of goods, firms in some fields (but not all) gain competitive advantage through employing strategies that include partnering, outsourcing, downsizing, networking, and the like. The precise relationship between network capital and other forms of capital is highly contextual and evolving. Nevertheless, works such as Castell's (1998) *The End of the Millennium* and Kelly's (1999) *New Rules for the New Economy* document network creation and maintenance as a critical aspect of competitive advantage in the contemporary era.

Although we have a reasonable understanding of the forces underpinning network formation, the effect of networks on the competition for advancement through education is poorly understood. For example, we do not properly understand the role of managers in constructing these networks with providers of tertiary education in contrasting fields (Lin 1999). Indeed, many accounts of the relationship between tertiary education and social class tend not to acknowledge that processes of class reproduction occur in different ways in different contexts. For example, on the basis of their study of elite labor markets, as they defined them,[2] Brown and Hesketh (2004) argue that employers prefer graduates from elite universities because, among other things, they are more likely to hold the required cultural capital. The authors argue that practices such as the outsourcing of the recruitment process mean that assessment centers are of increased importance. However, rather than disrupting the reproduction of elite groups, they argue that assessment centers have emerged mainly as a means of legitimating employers' decisions. Thus, attending elite universities is linked to subsequent advantage, almost irrespective of other factors. The authors conclude that many cases, much of the "social selection had taken place at earlier phases in the selection process" (p. 186).

Of course no single book can hope to provide answers to all our questions. Nevertheless, it is worth noting that like many studies, Brown and Hesketh's study does not properly acknowledge variations in the rules of the competition for advancement that arguably occur across time and space. For example, as is the case in New Zealand, the bulk of employment in the United Kingdom is with smaller employers, who are less likely to outsource their human resource functions. Similarly, as Brown and Hesketh recognize, a critical factor in determining graduate outcomes that cut across the kind of institution one attends is the kind of degree one obtains (Naylor et al. 2002). More generally, as shown in Chapter 3 of this book, research that has sought to test for the value of reputational capital is inconclusive. In Morley's (2007) research, for example, employers rated the reputation of the institution as one of the lowest considerations when recruiting new graduates. Another difficulty is that individuals do not enter institutions of tertiary education randomly. This means it is very difficult to attribute outcomes directly to institutional factors, such as its reputation or the quality of teaching. Indeed, although Brown and Hesketh's (2004) analysis enhances our understanding of some of the macro trends occurring in the graduate labor market and improves our understanding of how these trends relate to the creation of markets in higher education, they do not offer an account of how the strategies employed by groups and individuals to gain advantage might vary within fields.[3] For example, advancement in some fields may not involve attending elite institutions as the authors define them.

There are clearly winners and losers in the new network environment. For example, networks that are driven by certain forms of technological skill and expertise are concentrated in particular geographical locations. It is clear that in England, the development of networks is regional in nature, with some regions favored more than others (Potts 2002). Similarly, as Simmie and Sennett (1999) show, fields involving technological innovation levels are much higher in concentration in the areas running from East Anglia to the south coast than they are in northern regions. Evidence suggests that those areas located close to existing areas of innovation are best placed to enjoy spillover effects. In contrast, those areas found at some distance from innovative regions are at a relative disadvantage (Rey 2001). However, the concentration of innovative capacity in certain regions does not necessarily mean that inhabitants gain equally. Hudson's (2006) research demonstrates a significant correlation between the extent to which a region's economy has become knowledge based and its level of income inequality. The evidence is that as the proportion of people engaged in knowledge work increases, income inequality increases, suggesting that the benefits of highly paid employment are not shared by all.

Some innovation in New Zealand is also regional in nature; for example, the emerging creative film sector is situated primarily in Wellington, where filmmaker Peter Jackson is based. Although some have called for policies to redress regional variations in innovation (Potts 2002), the extent to which this strategy would have the desired effect has been questioned on the grounds that it may disrupt natural development processes (Martin 1999). When considering these comments, it is important to remember that there is no one dominant system of innovation and that it is likely that innovation is promoted in different ways in different fields. As noted earlier, some fields might develop systems of innovation that are largely confined to single nations, whereas others might be global in nature.

This book argues that state policy and the increasing significance of knowledge to innovation is contributing to the development of alliances between providers of tertiary education and firms. In terms of state policy, this has been motivated by changes in funding that have forced, for example, providers of tertiary education to become more entrepreneurial and by policies that are concentrating innovative capacity in relatively few institutions. As noted, the apparent concentration of innovative capacity is, in part, a pragmatic response from policy makers, who are facing up to the challenge of managing the strong demand for access to tertiary education while ensuring that financial support for research and development is not spread too thinly. It is also, arguably, in part a response from conservative elements within society to preserve the status of elite institutions, or those that provide tertiary level education to more privileged segments of society. In this respect, the preservation of universities' elite status is important to elite groups. This is because, in general, as Brown and Hesketh (2004) point out, these groups rely more heavily than other groups on academic qualifications to reproduce and legitimatize their status. Thus, by and large, elite institutions and elite groups in society have been supportive of the new policy environment because they benefit most from it. For example, the Labour-led Coalition in New Zealand is currently diverting funding away from institutions it perceives to be offering courses of low relevance to the economy and toward institutions offering courses deemed to be of high relevance, echoing developments occurring in England. In general, universities, and the middle-class students who predominate in them, have benefited from this move at the expense of nontraditional institutions. At the same time, the introduction of a new system of funding in tertiary education, in which enrolments will be limited, has arguably enhanced the ability of elite institutions in New Zealand to select the best students (see Chapter 8).

When considering these comments, it is important to remember that although elite universities appear to be best placed to create innovative ideas through research and development, innovation can emerge in all sectors of the economy, not just in those in which technological and scientific expertise is needed (Porter 1998), and network formation is likely to be important at every level of tertiary education. Blair's (2001) research shows that securing employment in the emerging creative sector in England depends on gaining and maintaining insider status and that the competition for a place in the sector is far from open, with those who possess social capital (in the form of social networks) best placed to succeed. This research suggests that initial employment in the industry is gained through kin- and community-based networks. Subsequently, professional networks and participation in teams is of greater importance (Blair 2001). One reason networks are of importance is that employers in this field typically use grapevine recruitment practices and employ teams of workers. One explanation for using network recruitment is that the skills needed to be effective in such contexts appear to be poorly measured by formal educational qualifications (this is explored further in Chapter 5).

The precise way network resources are of value will vary from field to field. However, in general terms, there are three reasons why we can expect networks to continue to be of significance in the recruitment process.

First, problems inherent within the education system mean that the qualifications on offer do not provide all of the information employers require to make effective recruitment decisions. This is partly because innovative ideas and related skills resist codification in forms such as qualifications (see Chapter 3).[4] Indeed, the fact that innovative knowledge is largely difficult to verify (and to transmit in asocial ways) is precisely the reason it provides competitive advantage and helps explain why employers continue to use network recruitment. In a related way, problems such as credential inflation and increased access to tertiary education are forcing individuals to find new ways of distinguishing themselves from their competitors.[5] Network resources are arguably of increased value at every stage of an individual's transition into tertiary education, as well as in their exit from it into employment. For example, networking resources can increase the likelihood that individuals select courses leading to qualifications that are desired by employers, and they can help individuals make effective links with employment. For policy makers and others, such networks are desirable where they help improve training decisions. Such improvements are badly needed because, at present, a lack of information about the value of some forms of training to employers means some people are effectively duped into gaining qualifications that have little utility in the labor market (see Chapter 4).

Second, innovation is fundamentally a social process. The basic idea underpinning this claim is that not all knowledge needed to generate innovative ideas can be embodied in a single individual. Moreover, much innovative knowledge is tacit in character and resists codification in such forms as educational credential. As a consequence, social relationships conducive to information sharing and skill transmission are often critical to innovation.

Third, changes in production and increased demand for graduate labor are facilitating changes in the skill mix required by employers. Here, there is evidence that employers are increasingly interested in the so-called soft skills and qualities, such as creativity, held by potential recruits. Such qualities and skills are difficult to evaluate through bureaucratic systems of recruitment (Rosenbaum 2002), raising questions about the links between tertiary education and work in some fields. In a similar vein, some new fields do not have well-established or direct pathways between training and employment. For example, holding a doctorate in biological science might signal technical proficiency in the relevant scientific skills needed to be effective in biotechnology. However, it says little about the ability of holders to use this knowledge in ways demanded in biotechnology, such as applying this technical knowledge in an entrepreneurial manner.

Although "third way"[6] administrations in England and New Zealand remain committed to the idea of equality of opportunity through education (Giddens 1998), policy changes designed to increase innovative capacity and encourage network formation are arguably undermining the realization of this commitment. One reason for this, which is explored more fully in this book, is that policy changes are concentrating research capacity in a relatively few, elite institutions, and they are forcing institutions (in New Zealand at least) to differentiate themselves from one another. In an era when the competition for advancement through education is increasingly intense, the concentration of innovative capacity in relatively few institutions improves the chances of students who attend such institutions, succeeding in the competition for advancement. Much of the rationale for increasing the participation in tertiary education of working-class students is that this will increase their ability to compete effectively with the already privileged in the race for positional advantage. It is designed to usurp advantage that is derived from forms of social closure found in tertiary education (Strathdee 2003). However, concentrating positional knowledge and other network resources within relatively few institutions may increase the chances that the elite will succeed in the race for advantage through tertiary education.

When considering this claim, it is important to remember that elite groups are not static, and neither are they homogenous. Rather, conceptions of elite are field specific, and the rules of the competition for advancement that give rise to the achievement of elite status are subject to change. Indeed, in some fields, globalization has shifted the balance of power toward individuals who are embedded in global networks and away from those embedded in domestic networks. More broadly, changes in the nature of production, such as those that result from the introduction of new technology or deviations in government policy, alter the rules of the competition and help bring new elites to power. Although it would be foolish to assert that all power now resides in networks, it can be argued that the emergence of economic, social, and technological change has shifted some power toward those who are embedded in the networks (Castells 1998). Networks are contributing in new ways to processes of class reformation. In this context, it is important to recognize that although, overall, those who attend elite universities currently earn more than those who acquire other forms of tertiary education, there is nothing predetermined or inevitable about this. Rather, changes in the labor market, and in the economy more broadly, mean that certain forms of capital have eroded in value while others have become commodified.

This book argues that in the current era, network creation—as a means of accessing innovative (social and technical) knowledge—and other resources have added currency. Although we can reasonably expect to find differences between fields, those institutions that are able to access and build networks in a variety of forms will be best placed to enjoy advantages in the contemporary period.

Building network capital requires creating what Florida (2002) calls "geographies of talent." For example, in a high-tech area of the economy, a combination of high levels of human capital, investment, sympathetic local policy makers, entrepreneurs, and philanthropists have come together to turn Oxfordshire in Britain from a "formerly sleepy, rural county . . . [into] one of Europe's leading centres of enterprise, innovation and knowledge" (Glasson 2003, 1). Underpinning this entrepreneurial activity is Oxford University. The 2001 Research Assessment Exercise revealed that Oxford had more 5/5* academic staff members than any other university (indicating that it had more academic staff members who had been judged to have reached national or international levels of excellence in research than any other university). Until recently, however, the commercialization of this research has been limited.

Toppling elite universities, such as Oxford, that have huge endowments of funds and are located in innovation-rich regions would be difficult. Nevertheless, institutions that are poorly embedded in networks can arguably improve their situation (in fields that use networks) through capacity building—strategic hiring and investment in network creation. For example, the formation of a research park in North Carolina over 40 years ago helped propel North Carolina from an economic backwater to an area of high innovation. In the area of biotechnology, the approach in North Carolina is seen as a model for the rest of the country. It was developed using the engineering and biomedical expertise of three local universities—Duke University, North Carolina State University, and the University of North Carolina–Chapel Hill—that are now considered world-class institutions. The formation of this cluster and the emergence of biomedical manufacturing plants did not occur spontaneously. Rather, they were "kicked into existence by government bodies" (Cooke 2002, 138) that supported their development through offering tax breaks and a supportive regulatory and planning environment. The role played by the North Carolina Biotechnology Center in coordinating the activities of government, industry, universities, and financiers was critical.

At the same time, Gough (2004) makes it clear that the creation of new networks is actually an act of network destruction. Accordingly, new networks and associations come to replace older networks and associations that have eroded in value; for example, the replacement of familial and community-based networks with new professional networks created and maintained by the State (Strathdee 2005a). Thus, a key difference in the contemporary period compared with earlier periods is the way the state is actively involved in network formation. In the United Kingdom, the Office of Science and Technology funds a variety of schemes designed to support the development of links between universities and businesses, through funding provided by the Department of Trade and Industry. Funding is not limited to elite universities—all institutions of higher education can submit bids for funding of innovative projects. Thus, although a key argument of this book is that funding changes are concentrating innovative capacity in relatively few institutions, current performance is no guarantee of future success. One only needs to look at the case of emerging research and development capacity in information and technology, which is growing rapidly in India and China, for evidence that globalization is altering the competition for advantage, leading to the formation of new groups of elite entrepreneurs (Saxenian 2002). Staying ahead in the race for advantage will require a continued commitment to capacity building through investing in network creation. All institutions of tertiary education have the capacity to build networks, though the precise nature of these will vary considerably within and between fields. In a similar vein, new fields are emerging, and considerable advantage can be derived by spanning boundaries between formerly discrete disciplines. In turn, in fields in which networks are of value, those institutions that can build networks with innovative firms will be better placed to provide their students with assets that confer advantage.

As noted, the changes outlined above are being buttressed by developments in tertiary education policy that are designed to make providers more entrepreneurial and institutions more accountable for the employment outcomes of their graduates. In New Zealand, for example, the Labour-led Coalition Government of 2005–2008 has followed the lead of New Labour in the United Kingdom by successfully arguing that not all universities should be funded to the same level. Instead, funds should flow to those institutions with the most research-active staff. At the same time, successive administrations have continued to reduce the proportion of public funding available relative to the number of students. This has created additional pressure on providers of higher education to gain external sources of funding. Some institutions have had great success. Of a total of 111 institutions in the United Kingdom, the top 15 institutions, led by Cranfield University and Imperial College, receive 50 percent of all industry support. Others, however, are less successful, with institutions in the lower half collectively receiving only 8 percent of the funding (Howells and Nedeva 2003). In turn, these developments are arguably cementing differences between providers of tertiary education, with some relying more heavily on teaching to generate income, and others on research. Although politicians and policy makers are right to emphasize the importance of qualifications to individual mobility and have introduced policies to encourage participation in tertiary education (as well as some that reduce it), these measures are unlikely to reduce social exclusion. Behind transitions into tertiary education and the labor market lie important social processes that play a vital role in securing advantage in the knowledge economy (Ball 2003).

Closely related to the idea that innovation results from social process is the idea that access to networks is becoming commodified. Authors like Castells (1998) are correct to highlight the emergence of the network society. However, many authors tend to characterize networks as porous and open. Although it can be argued that knowledge has the potential to operate as a public good and can, therefore, defy normal laws of scarcity that characterize commodity markets, policy developments in higher education are actually serving to commodify knowledge by limiting access to innovative knowledge to insiders. Gaining insider status depends on a number of factors, including an individual's ability to attend embedded institutions, as well as the possession of the social and cultural skills needed to cultivate networks. The precise mix of resources needed to advance depends on the rules of advancement present in the fields.

Chapter 2 begins detailing this book's argument that a better understanding of the value of contrasting forms of capital, conceptualizing competition for advancement as occurring in different ways in different fields, will improve our understanding of the state's role in structuring the rules of advancement through tertiary education. The next chapter reviews the relationships among networks, employment, and innovation. It builds on the insights of Hayek (1945) and others to argue that the formation of networks has implications for tertiary education provision. Chapter 4 looks at developments in tertiary education policy in New Zealand and the UK and describes how these contribute to the creation of geographies of talent. The following three chapters seek to test some of the conceptual arguments outlined earlier in the book. The first two draw on interview data that were gathered in a small number of interviews with managers. These data are used to explore the relationships among knowledge transfer, social networks, and innovation in screen production and biotechnology in Wellington and Auckland, respectively. Next, Chapter 7 turns attention to England, where knowledge transfer networks have been developed. Interview data, gathered in a small number of interviews with employers participating in knowledge transfer networks, are used to shed light on network processes. The final chapter provides a concluding discussion.

CHAPTER 2

Social Class, Tertiary Education, and Field Theory

Introduction

This chapter further explores the theoretical terrain on which this book is based. It begins by providing a very general description of contemporary debates concerning processes of social class reproduction ongoing in the labor market, some of which are mediated by the tertiary education system. The complex nature of the interaction that exists between individual-level variables, such as socioeconomic status and gender, and institutional variables, such as the programs on offer, as well as ongoing social and economic change, mean it is difficult, if not impossible, to provide a precise assessment of the relationships among labor markets, social class, and tertiary education. Despite our lack of understanding and the necessarily general nature of the discussion offered in this chapter, it is argued that a better knowledge of the role played by providers of tertiary education in constructing and using networks can improve our understanding of social class reproduction processes. Networks of various kinds work in diverse ways to structure and regulate social relationships, such as those based on social class (Lin 2001). An increased knowledge of such processes will enable us to understand better how providers of tertiary education are responding to the new policy environment.

As is widely acknowledged in the literature, it is important to understand how networks interact with other resources (or capitals). Achieving this necessitates that we examine interactions within fields (Bourdieu and Wacquant 1992). This is because different fields have different rules of advancement. For example, networks are likely to be a more valuable resource than educational qualifications in fields (conceived of as relatively discrete areas of the labor market) in which educational qualifications are poor signals of the skills and competencies demanded by employers. In such areas, employers may rely on other signals of the competency of potential recruits; for example, the reputation of the institutions where they have been trained or ascribed characteristics, such as the applicant's race or gender. This chapter argues that by providing a way to increase our understanding of the value of contrasting forms of capital, conceptualizing competition for advancement as occurring in different ways in different fields improves our understanding of the state's role in structuring the rules of advancement through tertiary education.

The following sections begin to build the case by describing the relationship between economic change and social class and by considering the role of social networks in this process. Although well rehearsed in the literature, this discussion is necessary because how this relationship is conceptualized has implications for how empirical evidence on the effect of tertiary education is interpreted and how network formation is understood. It also provides a point from which a better assessment of the effect of economic change on processes of social reproduction can be made. The next section of the chapter draws on Bourdieu's notions of field, habitus, and strategy to explore the value of different capitals operating in different contexts (Bourdieu and Wacquant 1992). As stated in Chapter 1, earlier works—for example, Brown and Hesketh (2004)—have improved our understanding of the relationship between tertiary education and the labor market. However, similar to the orthodox position in which it is maintained that all individuals who would like to work in the knowledge economy need to be highly educated and skilled (Kerr et al. 1973), much of this research tends to draw conclusions across fields.[1] One limitation of this approach is that it does not properly account for differences in the kinds of capital needed for advancement in different labor market contexts.

Although weaknesses remain, by theorizing that different kinds of capital will have different values or different effects in different social settings, Bourdieu and Wacquant's (1992) approach helps overcome such limitations. Understanding that different forms of capital have different values in different contexts helps lay the foundations for subsequent chapters, in which the role of the state in creating contrasting markets in tertiary education is explored.

Economic Change and Social Class

Over the last few decades, considerable effort has gone into assessing the costs and benefits of tertiary education to students. Although important variations to the rule exist, for most individuals, succeeding in tertiary education contributes to higher personal incomes, greater individual job security, and a greater sense of personal happiness, as British researchers (Brennan and Shah 2003) have shown.[2]

Such research has helped legitimate the technocratic-meritocratic perspective (Smetherham 2006). The technocratic-meritocratic perspective first emerged in the 1970s and describes the view that the introduction of new technologies and the need to integrate all individuals into the labor market according to their true ability was leading to meritocratic rules of advancement. As education became more important in the competition for advancement, the influence of social origins declined. The provision of comprehensive, state-funded forms of education and the development of national systems of assessment were all-important weapons in the attack on ascription.

It is widely argued that the rising significance of technological and scientific knowledge in modern economies, operating in conjunction with developments in, and the expansion of, tertiary education has helped separate origins from destinations. However, the continuation of inequality and ongoing change in the labor market meant that the struggle for meritocracy continues. For example, as shown later in this book, the desire to create meritocratic outcomes among students has led to the development of new forms of assessment (designed to assess competency more fairly) and new kinds of curricula (designed to better integrate people into the labor market). Similarly, because it is perceived by policy makers and their political masters to increase the likelihood that much—if not all—work in the future will be highly skilled and highly paid, investing in tertiary education contributes greatly to achieving meritocracy. Indeed, according to this theory, in technologically advanced countries, such as New Zealand and the United Kingdom, increases in the demand for skilled labor mean all workers, not just the elite, need to be highly educated and highly skilled. For such reasons, policy makers in the United Kingdom, New Zealand, and elsewhere have focused much attention on increasing access to tertiary education and on developing more motivating forms of curriculum and assessment.

In addition to evidence that demonstrates an earnings premium for those who hold qualifications, those who subscribe to this position point to changes in occupational structure—particularly the expansion of middle-class occupations—to argue that economic change is rendering the working class obsolete (Scott 1996). As a result, there is more room in middle and upper levels of the labor market, allowing those at the bottom opportunity to move up the ranks and claim a place in the swelling middle classes. Providing those currently at the bottom of the labor market with the means to move up the occupational scale depends on the continued expansion of tertiary education, so that these people can gain the knowledge and skills, and ultimately the qualifications, needed to win a place in the growing knowledge economy.

Further arguments supporting the idea that economic and social change and intervention by the state are driving up demand for workers who hold different kinds of skills compared with those in demand in the past can be garnered from changes that are perceived to have occurred in the way production is organized in advanced economies. Although there are other ways to interpret these shifts, the underpinning logic is that in Western nations, neoliberal, decentralized, service-based economies driven by principles of flexible accumulation are replacing the once dominant, socially democratic, centralized commodity-producing economies of the 1960s and early 1970s. At a very general level, the newly emerging forms of social organization are thought to be part of a broader shift from Fordist systems of production to post-Fordist systems of production. The emergence of post-Fordism in the mid 1970s is also thought to have led to important changes in how people work. The key changes include an increased emphasis on teamwork, a blurring between mental and manual work, and an increased need for workplace flexibility—a term that is used to capture behaviors and attitudes, including openness to change and a "can do" approach to work. Related developments include the formation of so-called flat hierarchies within workplaces and growth in the significance of network production.

All of these developments hold out the possibility of progressive change in a number of areas, including education, democracy, and employment. For example, in the area of employment, the view is that the creation of flat hierarchies in workplaces is increasing the ability of workers to have a say in how their work is organized (Lloyd and Payne 2002). In the past, class antagonisms were clearly visible in the workplace and involved conflict between the owners of the means of production (the ruling or capitalist class) and workers (the working class). One manifestation of this conflict was growth in trade unions, which were thought to represent the interests of the working class as they struggled for a greater share of their labor's produce. However, ongoing social and economic change, working in concert with the gradual emergence of a meritocracy, has effectively neutralized such conflict and heralded a new era in which workplaces are increasingly characterized by new forms of democracy. In the process, the old discourse of unions—class struggle and conflict—has been replaced with a newer discourse of worker participation, empowerment, partnership, consultation, stakeholder economy, and networks. Workers and the owners of production are no longer seen to be at war with each other. Rather, they must work together to bring about innovation and, hence, retain competitive advantage (Brown and Lauder 2001). For example, facilitating greater cooperation between workers and management in individual workplaces is an innovation that can improve productivity by allowing workers increased input into production processes. Thus, the need for greater cooperation (and the resulting increase in workplace democracy) and participation in networks is underwritten by changes in production, which are encouraging people at all levels of society to work together to add value to their enterprises and to resist competition from foreign producers and foreign regions.[3] In related developments, the emergence of global networks, clusters of industries, and regional, if not global, systems of learning and assessment means that local resources no longer play the critical role they once did in identity formation. For example, as the work of Florida (2002) has popularized, one's place of origin no longer exerts the influence it once did over where workers live and work; neither does it exert the influence it once had on the identities they develop. Indeed, Florida argues, in the case of workers in the rapidly growing creative fields, although cities and regions continue to be organized along class lines, we are also simultaneously witnessing a large-scale re-sorting of people among cities and regions nationwide. As part of this restructuring, we are seeing different groups gain advantage. For example, evidence suggests that women aged in their 20s who work full time and live in cities such as New York, Chicago, Boston, Minneapolis, and Dallas now earn more than men of the same age. In Dallas, for example, women earn 20 percent more than men; in New York, the figure is 17 percent (Breitman 2007).

Along with low barriers to entry for innovative companies, centers can become home to members of the creative class by becoming plug-and-play communities. By providing easy access to creative activities, such communities allow creative workers further opportunities to build their own identities and enjoy a high quality of life. To illustrate his argument, Florida provides examples of creative regions throughout the world, including Wellington in New Zealand, which he specifically identifies as an area of creativity linked to the screen production industry (see Chapter 5).

In a similar refrain, it is argued that economic and social change, along with the availability of new knowledge through networks and education, has reduced the significance of local contexts and families on identity formation. Indeed, the working-class identities that were created during the mid-1900s are no longer relevant, and new sources of identity formation are now available, for example, by participating in new social networks and education (Beck 1992). Indeed, geographers have for some time argued that space is no longer necessarily local; rather, processes of globalization mean that space is increasingly global, and the significance of locale has declined (Amin 2004).

Questioning the Technocratic-Meritocratic Perspective

Numerous works, such as those cited later, have critiqued the view that society is becoming more open and that greater investment in education and training has played a critical role in this outcome.[4] Some critiques of the official view, or what has been referred to as the "training gospel" (Jordan and Strathdee 2001) point out that simply increasing participation in tertiary education will not, in itself, necessarily realize the promised social and economic benefits. One source of argument is that changing skill demands have not kept pace with the production of credentials. This has resulted in a good deal of overeducation. Overeducation can result in a situation in which the qualifications and training gained in tertiary education are in excess of those required to enter the labor market (Smetherham 2006; Livingstone 1999). For some, overeducation, or "credential inflation," is also an outcome of strategies employed by elite groups (such as those middle-class groups whose advantage is linked to credentialism) to rig the competition for advancement through education in their own favor (Brown 2000). One way elite groups achieve this is by continually bidding up the qualifications needed to enter the most desirable positions in the labor market. This happens at the expense of those less well equipped to stay on and achieve such qualifications. The state supports this process by establishing and policing the rules of the positional competition for advancement. For example, by expanding the provision of tertiary education, and further developing credentialism, tertiary education reform has helped reproduce, rather than challenge, the advantage enjoyed by existing social class groups.

In support of this position, researchers report that, after controlling for education, the effect of class origin on subsequent destinations is substantial and shows little sign in Britain (Breen 2004) or Ireland (Whelan and Layte 2002) of having diminished over time. Similarly, Corak (2004) shows that the increase in student enrolments in university education during the contemporary period are associated with reduced—not increased—social mobility. Halsey (1993) and Wolf (2002) also argue that despite expansion of higher education, social class inequalities in the United Kingdom, measured in relative terms, have either remained stable for the last three generations or increased as Britain has become less meritocratic. New Labour's own evidence indicates that "young people from professional backgrounds are over five times more likely to enter higher education than those from unskilled backgrounds" (Department for Education and Skills 2003, 17). Evidence drawn from a number of countries suggests that much of this inequality is transmitted between generations.[5] For example, in a cross-national comparison of generational earning mobility in North America and Europe, Corak (2004) estimates that up to 50 percent of the parental income advantage is passed on to their respective children.[6] Such findings suggest that educational reform has been unable to substantially change the rules of advancement in ways that limit the abilities of existing elites to rank highly in the race for advancement through education.

It is important to recognize, however, that inequality is not transmitted equally or uniformly across generations. For example, on the basis of their reading of the available evidence on the U.S. case, Bowles and Gintis (2002) argue that general data on the level of wealth transmitted between generations can hide important differences between subgroups. They identify intergenerational wealth transmission as being strongest in the lowest- and highest-income groups, such that the children of the very wealthy and the very poor are most likely to follow in their parents' footsteps. Bowels and Gintis (2002) also identify a distinct racial pattern, with successful blacks finding it more difficult to pass this advantage on to their children than similar groups of white families. To extend Bowles and Ginitis' (2002) arguments, it can be asserted that intergenerational wealth transmission is more likely to occur in some fields; for example, those that are based on family ownership of capital, such as small businesses, than in others, where the rules of advancement are meritocartic.

As noted above, recruitment into tertiary education has come into focus as policy makers and politicians further advance the technological meritocratic perspective and face up to the issue of social inequality. One idea is that it is inequality in access to tertiary education that is fueling inequality between adults. For example, the Schwartz Report (Admissions to Higher Education Review 2005) on access to a university education in England is underpinned by a desire to improve the participation of students from nondominant backgrounds in tertiary education, particularly in higher-status institutions. The belief is that, to date, access to high-status institutions has been hampered by the exclusionary tactics of elites, who tend to enrol those who come from privileged backgrounds rather than those from the wider population. According to the Schwartz Report, one reason for this in the case of English universities is their tendency to rely heavily on A-level results as a basis for selecting students. However, the Schwartz Report argues this strategy is flawed because the relationship between success in A-levels and subsequent success in some tertiary level discipline areas is, at best, flimsy.

As the Schwartz Report shows, the massive expansion in the provision of tertiary education accompanying the introduction of markets into the sector has seen participation of those from all socioeconomic (as a proxy for social class) backgrounds increase. However, as the Schwartz Report noted and other research has confirmed, although young people from all social backgrounds have increased their level of participation at university, growth in the proportion of young people from low-socioeconomic groups has been lower than that found for higher-socioeconomic groups (Galindo-Rueda et al. 2004). Related research shows a heightened connection between family income and attainment in higher education. The expansion in higher education that has occurred between the 1970s and late 1990s primarily benefited those who were already relatively well-off. For example, Blanden and colleagues (2005) show that young people aged 16 years from all social backgrounds are benefiting from increased rates of participation, with those from the poorest 20 percent of households in Britain increasing their rate of participation from 21 percent in 1974 to 61 percent in the late 1990s. The comparable figures for the richest 20 percent were 45 to 86 percent, respectively. More critically, however, although those from low-socioeconomic backgrounds have kept pace with their higher-socioeconomic peers in terms of entering higher education, they have not kept pace in terms of degree completions. Between 1981 and the late 1990s, those students at age 23 from the poorest 20 percent of households only managed to increase their degree completions from 6 percent to 9 percent. In contrast, those of the same age from the richest 20 percent of households increased completions from 14 percent in 1974 to 46 percent in the late 1990s (Blanden et al. 2005).

Recent research from Australia produced similar findings. Students from lower-socioeconomic backgrounds were half as likely to participate in higher education than were those from medium- and higher-socioeconomic backgrounds (James 2002). Similarly, smaller-scale research from New Zealand also shows that although between 1983 and 2001 students from middle-socioeconomic backgrounds increased their level of participation at university, the participation of students from low-socioeconomic backgrounds has remained essentially unchanged. However, during the same period, the advantage enjoyed by students from high-socioeconomic backgrounds appears to have declined, although they remain comparatively advantaged (Strathdee and Hughes 2006).

Although students from lower-socioeconomic backgrounds tend not to make use of tertiary education to the same extent as those from higher-socioeconomic backgrounds, whether or not this reflects any inequality of access remains a subject of debate. Some researchers have argued that although higher-socioeconomic groups tend to dominate higher education, almost all of those who are qualified to attend higher education do so, and there is no evidence of large-scale inequity in admissions to higher education (Gorard 2005). Similarly, in New Zealand, Hughes and Pearce (2003) found that students from low-decile (low-socioeconomic) schools lacked the entry qualifications needed to gain admission into tertiary education. For this reason, the authors argued that changes to the system of student tuition fees to increase the participation of students from low-decile schools were unlikely to have much effect. Given these findings, these researchers have argued that attention needs to be focused on addressing those factors that cause students from low-socioeconomic backgrounds to lack the required qualifications in the first place. Indeed, Gorard (2005) suggests that effort should be exerted toward addressing such aspects of education such as differences in student retention rates and their attainment at sixth-form level (through, e.g., interventions designed to increase student motivation to learn), as well as increasing access to preschool education. In this respect, recent developments in education indicate that policy makers perceive there to be a problem motivating some young people from low-socioeconomic backgrounds to gain the qualifications they need to gain employment. To rectify this deficiency, the state is attempting to remake identities. This is not the place to provide an extended review of the arguments. However, it is useful to note that an important aim of policies in the area is to extinguish those attitudes and behaviors that limit the participation and achievement of those who have traditionally fared poorly in the education system. For example, in the United Kingdom, the Aimhigher campaign, which aims to encourage young people to think about the benefits and opportunities provided by higher education, is designed to motivate students who have traditionally not gone to university to do so (Higher Education Funding Council 2007). Similarly, through the Connexions Service, young people can gain financial support while they study, as well as financial bonuses for completing their studies (Department for Education and Employment 2000). An aspect of the argument advanced is that the lower levels of success attained by some social groups reflects both an unwillingness to adopt the new learning cultures needed to gain the qualifications required to enter and then to succeed in tertiary education, and a reluctance on the part of institutions to remove barriers to their participation. Meeting the challenges thrown up by economic and social changes makes it imperative that identities rendered obsolete by economic and social change be modernized. For example, according to the official view, it is no longer sufficient to reject the possibility of participating in tertiary education and training and remain in secure, well-paid employment. Rather, all individuals need to adopt proactive attitudes toward engaging in ongoing learning and skill development.

To aid in this process of remaking learner identities, new educational technologies continue to be developed. For example, in high schools, modularization of the curriculum, the adoption of standards-based assessment, and the construction of qualifications frameworks (such as the New Zealand Qualifications Framework) have, in part, been developed to facilitate the creation of lifelong learning identities and to turn "Willis" lads into earoles (or students who conform to the culture of the school; see Chapter 4). To a large extent, then, efforts to promote greater meritocracy are also exercises in identity reformation through reform.

The fact that the state continues its attempts to remake identity suggests, first, that economic and social change mean that the state's work in this area is ongoing, and second, that its policy interventions are not completely effective. In evidence of the latter, the notion that students have been emancipated by the erosion of traditional cultural codes by the expansion of choice in tertiary education has been challenged on the grounds that such claims misinterpret the way in which people experience tertiary education. For example, critics point out that significant groups of people have not adopted prolearning identities. Rather than converting those who have obsolete identities as learners into active lifelong learners, tertiary education reforms have been rejected by segments of working-class students, possibly on the grounds that these reforms do not meet their needs (Archer and Hutchings 2000; Mizen 1995), One expression of this (as evidenced in the Schwartz Report) is the failure of tertiary institutions to attract sufficient numbers of young people from nondominant backgrounds and the tendency for a disproportionate number of young people from such backgrounds to not complete their qualifications. However, although measures such as Aimhigher and the removal of institutional barriers to participation are intended to make the system more motivating and inclusive, it can be argued that the relatively low levels of participation of students from low-socioeconomic backgrounds reflect negative learner identities that were established early in their lives. For example, Gorard and Rees' (2002) quantitative research found that characteristics established very early in an individual's life, such as family background, predict later lifelong learning trajectories with 75 percent accuracy. Adding additional variables representing initial schooling increased the accuracy of their predictions to 90 percent. Those individuals who do participate in postcompulsory education are thus, "heavily patterned by 'pre-adult' social, geographic and historical factors such as socio-economic status, year of birth and type of school attended that are related to subsequent participation" (Gorard and Rees 2002, 7). For Mizen (1995), however, the students' relatively low levels of engagement in tertiary education simply reflect the antipathy they hold in relation to education generally. Addressing the obvious barriers to participation such as cost, time, and location will not rectify this problem.

Another explanation offered in the literature for this antipathy is that by observing labor market transactions, people learn that educational credentials are not always needed for advancement into the labor market, and so perceive little benefit from staying in education and training (Rosenbaum 2002); that is, the anticipated rewards are not sufficient to motivate them to achieve the qualifications they need to gain entry into tertiary education programs. Indeed, increasing retention and attainment (with a view to increasing participation in higher education) will only pay dividends if the rules of the competition for advancement in the fields in which students eventually compete actually involve education and training and the use of qualifications.

It is also important to remember that reducing social class differences involves much more than simply increasing participation and achievement through remaking identities and related measures. As critics of the official view point out, access to different kinds of higher education remains differentiated according to social class, and simply increasing the participation of students from low-socioeconomic backgrounds in low-status programs will do little to erase social class divisions (Leathwood 2004).

Participation in Tertiary Education

At this juncture, it is useful to review some of the research that has explored the relationship between participation in tertiary education and the labor market. Recent research into the relationship between participation in tertiary education and labor market outcomes has highlighted the difficulties of conducting research in the area. In this respect, evidence shows that returns from succeeding in tertiary education are mediated by a complex mix of factors including the course undertaken, the institution where it was held, the gender of graduates, and the field in which graduates work. When considering possible relationships, it is important to remember that income is not the only measure of success and that fields define success in a diversity of ways. Nevertheless, if the bonds between elite institutions and elite forms of employment remain strong as some suggest (Brown and Scase 1994; Scott 1995), it is reasonable to expect to see this reflected in quantitative research.

In relation to this, Brown and Hesketh's (2004) study reports that candidates from Oxford University applying for blue chip jobs with one leading employer were 29 times more likely to be appointed than those applying from a "new" (post-1992) university. In other research that looked at the relationship between the kind of institution attended and income,

Conlon and Chevalier (2003) report an earnings advantage of Oxbridge graduates of almost 8 percent over those who attended a new university. Those attending former polytechnics experienced an earnings penalty of almost 4 percent compared with those attending old universities. However, Blundell and colleagues (2005) found in their study of the National Child Development Survey in the United Kingdom that in general, there is much variation in the returns to education across individuals who held the same educational qualifications.

It has also been reported that graduates from elite institutions fare best in terms of their ability to secure a place in fast-track graduate trainee programs. In one research project, graduates from elite institutions were found to be four times more likely than those from low-ranking institutions, or those institutions found at the lower end of the published league tables in terms of research and teaching, to have gained a place in fast-track graduate trainee programs (Smetherham 2006).

It is worth noting the results of Smetherham's (2006) research, as they provide a useful assessment of the relationship between background factors and success. In terms of gender, Smetherham's study found that males with first-class degrees earned more than females with the same qualification for up to five years after entering the labor market. In part, this finding reflects a greater tendency on the part of females to enter public sector occupations, particularly modern professional and clerical/intermediate occupations, which include teaching and nursing. Smetherham's research also found that graduates from elite and midranked institutions also fared the best in terms of accessing jobs for which formal qualifications were required. In such institutions, less than one fifth of graduates reported working in jobs in which qualifications were not a formal requirement. In contrast, in excess of one half of graduates from low-ranking institutions reported working in jobs where degrees were not a formal requirement. Graduates from arts/humanities and the natural sciences with firsts fared significantly worse than those from other discipline areas, both in terms of income and skill utilization (Smetherham 2006).

In terms of income, 35 percent of recent graduates from elite institutions with first-class degrees were earning over £25,000 a year, compared with just 15 percent of those from low-ranking institutions. After five years in the labor market, almost 50 percent of those who gained firsts from low-ranking institutions were earning less than £19,999 a year, compared with less than 20 percent of those from elite universities. At the other end of the labor market, about one third of those from elite institutions were earning over £40,000 a year, compared with just 5 percent of those from low-ranking institutions (Smetherham 2006). In a similar vein, Smith and colleagues (2000) found that *ceteris paribus*, students from poorer backgrounds who graduate from higher education institutions have a lower chance than those from wealthier backgrounds of being employed in graduate occupations.[7] Finally, Chevalier and Conlon (2003) estimate that after accounting for background factors such as socioeconomic status, when compared with graduating from a modern university, graduating from a Russell group university[8] adds a wage premium of between 0 and 6 percent for men and up to 2.5 percent for women.

Similar findings are reported in studies conducted in the United States. For example, on the basis of their examination of the National Longitudinal Study of the High School Class of 1972, Brewer et al. (1999) found, after controlling for selection effects, strong evidence of a significant economic return on attending an elite private institution. Similarly, Zhang (2005) found that the common wisdom that it pays to attend high-quality institutions to be quite robust over an array of measures of college quality. However, the evidence is inconclusive, with researchers also finding no or little advantage to attending elite tertiary institutions. For example, after accounting for ability, researchers from the United States found no earnings advantage gained by those who attended selective colleges[9] (Dale and Krueger 2002). In this respect, James et al. (1989) argued on the basis of their research that although sending one's child to Harvard appeared to be a good financial investment, sending them to a local state university to major in engineering and then encouraging them to gain a good grade point average was an even better investment. What mattered most in James et al.'s research was what one studied, not where one studied, suggesting that field of study has an important effect on outcome.

One difficulty with this kind of research is that the graduate labor market is complex and the source of advantage and disadvantage remains unclear. An important limitation is that the quantitative models employed are not able to properly account for background variables that affect outcomes; another is that the data employed have not always been gathered to address the specific research questions at hand. This means that poor proxies of quality and weak measures of effectiveness have been employed in these studies. Moreover, although educational researchers have identified an effect of college quality in the United States and England, Zhang (2005) argues that researchers may have a difficult time in identifying the sources of the advantages those high-quality institutions provide. Possible explanations identified by Zhang include peer effects, better resources, higher level of engagement, sorting effects, and possibly favoritism.[10]

As noted, one limitation of the official view is that it makes claims about the value of tertiary education across all fields. Similarly, researchers drawing conclusions about the relationship between participation in tertiary education and labor market outcomes at a general level may ignore the importance of differences that exist across time and space. For example, large variations exist between nations in terms of the mix of private and public and selective and nonselective tertiary institutions. Such differences are likely to influence the extent to which tertiary education influences graduate outcomes. For example, in a homogenous tertiary education system, we can reasonably expect greater uniformity in graduate outcomes than among those produced by a heterogeneous system. Differences in degree programs, differences in the background characteristics of students (such as their socioeconomic status), and a myriad of other factors make comparing the effect of different institutions on student outcomes highly problematic.

Capitals and Fields

The argument so far in this chapter is that access to tertiary education remains structured by social class (though the reasons for this seem to lie in earlier levels of the schooling system), whereas the outcomes from participating in tertiary education are influenced by a range of factors including the field in which the graduate works (James et al. 1989), as well as background factors such as gender and socioeconomic background.

Our understanding of the way in which competition for advancement is organized can be enhanced by conceiving of the competition as occurring in different ways in different fields. Briefly reviewing the work of Bourdieu (1997), Coleman (1988), and others such as Erikson (1996) provides a way to start this task.

One benefit of conceiving of the competition for advancement as occurring in different ways within different fields is that it provides a better way to understand the results of research on social class reproduction. Another benefit is that it allows us to appreciate more the kinds of assets that may or may not be available through social ties, both within and between different fields. In turn, such an analysis provides a better way of conceptualizing the kinds of assets individuals need to possess if they are to access network resources and to understand how state intervention might alter the mix of resources individuals need to hold if they are to advance.

Bourdieu and Wacquant (1992) argue that children pick up a deeply ingrained, largely unconscious orientation, or habitus, from their social environment. Habitus binds individuals to groups and exerts an important influence on the reproduction of social class groups. This is because elite groups have better access to dominant forms of habitus, and these groups are better able to assert their status. Habitus, in the form of culture, is also a form of cultural capital. Those with superior habitus have more cultural capital and are best placed to gain qualifications, which are used to legitimize elite status.

In their model of social class reproduction, Bourdieu and Wacquant (1992) argue that the transmission of habitus through socialization is largely immutable. Although people can attempt to adopt the habitus of elite groups, they cannot do so in a convincing way. For example, the nouveaux riche can accumulate goods symbolic of elite status, such as fine art and grand houses (though both forms of capital only have currency with segments of the elite), and they can increase the access their children have to elite cultural capital by, for instance, sending them to elite schools. However, according to Bourdieu Wacquant (1992) at least, they can never join the ranks of the aristocracy because they will always lack the required cultural capital, indicating that family background has an important, if not determining, influence over a person's destination and reduces the possibility of social mobility.[11]

Although Bourdieu and Wacquant's (1992) account offers insights, in practice, it has proven difficult to identify a hierarchy of culture, and it seems that there is no one kind of cultural capital that is of importance in all fields. In the field of higher education, institutionalized cultural capital (such as educational qualifications) is more important than it is in the field of business, where economic capital is arguably more highly valued.[12] Both professors and business leaders belong to elite groups, but the capitals required to achieve this status differ. This relatively simple illustration highlights a problem with much research into social class and related debates—namely, that the connection between qualifications (and other forms of capital) and advancement is field specific.

Empirical research has also shown that high-status individuals are more likely than low-status individuals to consume elite culture, but there is no unifying feature to this consumption. In other words, "there is no one kind of taste profile that advantaged people share" (Erikson 1996, 219). Indeed, some economic elites tend to reject high-brow culture, dismissing it as an extravagance. Also in contrast to Bourdieu Wacquant (1992), a further argument noted by Erikson is that individuals can develop proficiency in a variety of cultures and therefore can be effective across fields. In her 1996 study, Erikson found that in one field (as she defined it—the contract security industry in Toronto, Canada), cultural variety was linked to advantage. Working effectively in this field required individuals to be at ease across boundaries. Elites in this field tended to have diverse networks and were cultural omnivores. That is, they engaged in the consumption of more forms of culture than subordinate groups, and only a minority of these elites shared any particular forms of culture or consumption. It was their ability to operate in different contexts that led to their advantage. Thus, rather than being an immutable result of early socialization, by Erickson's account, habitus can be developed through participating in different or culturally alien networks. As described later in this book, one reason developing this competency leads to advantage is that it increases the ability of individuals to become boundary spanners, or those who can work effectively across two or more fields.

In spite of the weaknesses identified by Erikson (1996), the approach taken by Bourdieu and Wacquant (1992) provides a method to understand better the competition for advancement through education. In Bourdieu and Wacquant's (1992) model of class reproduction, cultural capital operates in conjunction with economic and social capital in processes of social class reproduction. Economic capital refers to financial assets and finds expression in general terms as property rights, whereas social capital refers to resources that can be mobilized through network membership. It is useful to consider social capital in greater depth.

Debate continues, however, about how to define the concept of social capital, how it is operationalized, and how to measure it. Many variations of the concept are based on the basic premise that social capital is "not what you know, it's who you know" (Woolcock and Narayan 2000, 225). For most authors, social capital is generally defined in terms of networks, norms, and trust. In various configurations, and in different ways in different settings, social capital confers advantage. For Bourdieu (1997) and Coleman (1988), social capital is embedded in concrete social relations that permit individuals and groups to access resources. Social capital is context specific and can be built up through repeated exchanges in which process-based trust is established (Zucker 1986). Thus, social capital can be a product of investment strategies, such as the cultivation of social networks, employed by individuals, groups, and the state. Social capital can take a variety of forms, such as bonding, bridging, and linking. As Putnam puts it, "bonding social capital constitutes a kind of sociological glue, whereas bridging social capital provides a sociological WD-40" (Putnam 2000, 23). Linking social capital refers to networks and institutionalized relationships between unequal agents (Szreter 1998), which involves interactions between actors horizontally across the social structure.

It is important to note that the effects of these various forms of capital may be positive or negative, and the presence of networks says little about their effects. For example, bonding social capital might help individuals deal with poverty through allowing them increased access to community support. However, it might also reduce options in the labor market by limiting the formation of bridging social capital. When assessing whether or not networks should be considered to be social capital, knowledge of the assets gained by membership of networks, as well as the context in which they are deployed, is critical.

Although Tilly (1998) does not use the concept of social capital, his ideas clearly resonate with it. Tilly developed the idea of opportunity hoarding to explain how certain groups structure the competition for advancement to their advantage. Opportunity hoarding occurs when groups do not have the same access to resources that confer advantage; for example, when newly arrived immigrants take control over a valuable resource, such as information about employment opportunities. Such groups hoard access to these opportunities by only sharing employment related information (e.g., information about actual vacancies and information about the skills needed to be effective in the workplace) with those in their social networks. Employers support the formation of such opportunity structures through relying on grapevine recruitment and on the patterns of obligation that develop. Related case study research shows one reason employers do this is that it allows them to exert greater control over recruitment and over the labor process (Cranford 2005).[13]

Advantage that is derived from the deployment of different forms of capital in different fields is further developed by Bourdieu and Wacquant (1992) through their notion of strategy. In Bourdieu's earlier works (Bourdieu and Passeron 1977), his conception of strategy is dependent on habitus. As noted, habitus is a result of socialization, which engenders in people attitudes and dispositions beneath the level of consciousness. The concept of strategy is typically conceived as a specific orientation of practice. Usually, this is based around notions of conscious calculation or deliberate action. The adoption of practices that give advantage can be conceptualized as position taking. However, in later works, in a position that resonates with neo-Weberian sociologies' notions of rigging and ranking (Brown 2000), Bourdieu extends the possibility that social capital can be the result of deliberate investment strategies in network creation. Thus, "the network of relationships is the product of investment strategies, individual or collective, consciously or unconsciously aimed at establishing or reproducing social relationships that are directly usable" (Bourdieu 1997, 52).

An important additional contribution made by Bourdieu and Wacquant (1992) is to give emphasis to the ways in which different resources (or capitals) have different values in different settings (or fields). Bourdieu argues that social formations, and the institutions these give rise to, are structured around a complex ensemble of social fields. Fields are structured as a "network, or configuration, of objective relations" among positions (Bourdieu and Wacquant 1992, 97).

The idea that different capitals have different values in different fields is helpful because it allows us to understand that the rules of competition for advancement through tertiary education differ in contrasting settings and contexts. Fields vary by the relative importance ascribed to various capitals, but individuals compete on the basis of some combination of these. From this perspective, class differences are largely the result of the solutions elite and nonelite groups devise in the face of recurrent problems and challenges to their status. These challenges typically center on control of material, cultural, positional, and symbolic resources (Emirbayer 1997). However, although it is possible to describe the contest at a macro level, it is important to recognize that the way the contest is played out is best understood locally. A key reason for this is that although changes in the overall framework can occur at a macro level (for example, the shift from social democracy to neoliberalism), the precise nature of the resources needed for advancement can only be understood locally (although they will have a global dimension in some fields).

To return to Bourdieu, fields are structured in a hierarchal fashion, with different forms of capital being of greater or lesser value in different fields. This argument holds relevance for our understanding of the competition for advancement through tertiary education because it alerts us to the fact that the rules do differ. However, whatever the field, some combination of economic, cultural, and social—as well as other forms—of capital, is required for advancement.[14] So, for example, given the importance of personal referrals to their work, real estate agents arguably rely more heavily on social capital than on human capital to be successful, whereas academics rely more heavily on human capital than on economic capital to be successful.[15] However, even these examples are problematic because conceptions of what makes up human capital vary from field to field—in some fields, physical strength is a form of human capital, but in others, it is not. In addition, conceptualizations of "elite" are field specific. This means that what counts as elite status, the resources needed to achieve this status, and the rules that govern the competition for advancement are best understood as occurring in different ways in different fields. From this perspective, the valuing of elite forms of cultural capital in schools and the use of educational qualifications to reward those who hold elite culture is an exercise in boundary formation and maintenance. Battles over what knowledge is valued in schools, what rituals and other cultural practices are held in high esteem, and how competency is assessed and signaled are all in part struggles over boundaries (or class struggles). Changes in schooling, such as the introduction of standards-based assessment, have been driven by a desire to redraw the boundaries so that they are more inclusive (see Chapter 4). However, elites within fields attempt to preserve their status by developing new boundaries as threats to their status emerge (Lamont and Molnar 2002). For example, the strategy of "professionalization" can be seen as an attempt to introduce new forms of closure (Murphy 1984).

Although conceptualizing the competition for advancement as conducted in different ways within different fields provides one way to theorize the relationship between tertiary education, the labor market, and advancement, Bourdieu did not give a strong indication about how the boundaries between fields should be drawn. One solution advanced in the literature as a method of coping with problems identifying boundaries between fields is to let participants who are engaged in competition drive the boundaries of fields. This means that determining the boundaries between fields must remain a subject of research and testing.

According to DiMaggio and Powell (1983), the structure of a field is largely a result of patterns of relations between—and choices made by—individuals. These interactions remain structured by the rules in operation in particular fields as well as the nature of the boundaries between fields. Thus, fields are organized around a range of significant actors and their relations, and as Scott (1995, 135) points out, they are "coterminous with the application of a distinctive complex of institutional rules." This means we should define fields on the grounds of the interactions found, rather than on the basis of abstract principles established before research and testing. However, one danger of adopting this approach is that fields become known for their effects, and it is tempting to proliferate invisible fields to explain that which we are unable to explain otherwise. Another solution, which is the approach adopted in the case studies included in this book, is to draw on existing knowledge to define fields ahead of empirical research and testing. For example, some define higher education as a distinct field (Naidoo 2003).[16] On the one hand, this seems sensible, as institutions of higher education engage in similar activities. For example, they all award degree qualifications and undertake research. However, on the other hand, as the evidence presented above suggests, the field of higher education is diverse. For example, the returns from engaging in higher education differ from institution to institution and from program to program. Similarly, the entry requirements into universities also differ markedly, with some institutions being highly selective and others taking virtually all comers (Gorard 2005). The fact that there are differences in the tertiary education systems offered in nations across the world compounds the problem (Bergh and Fink 2005). A further complexity exists in that it is likely that different mixes of capitals can lead to advantage in the same fields. For example, in some fields, those who lack strength in some forms of capital can still advance because they are well endowed in other forms of capital.

In addition to problems with delineating fields, we know relatively little about the conditions in which various capital interacts within fields, as researchers have constructed them. Indeed, although recent work has improved the situation (Ball 2003), most conventional analyses of social-class reproduction are partial because they focus on cultural and financial capital as the key determinants of educational success and future placement in the labor market while neglecting the role of social capital (Wong and Salaff 1998) and other forms of capital. Moreover, as noted, understanding the interaction between various capitals is made more challenging by social and economic changes that are forcing a reconfiguration of their relative values. Thus, researching social processes requires a commitment to ongoing investigation within fields.

In characterizing the formation of competition for advancement in this way, it is important to remember that both elite and nonelite groups do not necessarily work in unison to advance their collective interests. Rather, as Bearman (1993) reminds us, differences exist in the actions of elite groups. Elite groups are usually in competition with each other, and although they may unite around specific issues, individual activities are not always—or necessarily—motivated by membership of a category but, rather, by the actual social relationships within which individuals are located. Moreover, as implied above, even within relatively discrete elite groups, there will be differences between the kinds of capitals individuals can muster.

Despite weaknesses, such as those noted above, Bourdieu and Wacquant's (1992) conception of fields provides a way to understand how different kinds of capital can be combined in different ways to produce patterns of advantage and disadvantage. One way to theorize these issues in relation to social networks is to see them as providing a conduit through which assets of various kinds flow. Social and economic changes may alter the mix of resources needed to secure advantage; however, their value continues, in different forms in different fields, across time and space. For example, the "obsolescence of old social networks and family influences on the transitions to work" has been associated with an increasing importance of formal education (Fevre 2000, 107). However, as argued, far from eroding the importance of education, the introduction of user pays in higher education may have reasserted the importance of the intergenerational transfer of financial resources (Ahier and Moore 1999). In this respect, Galindo-Rueda et al. (2004), who concentrate their analysis on the 1990s, found that although growth in participation of students from low-socioeconomic backgrounds was high, growth in the participation of students from high-socioeconomic backgrounds was higher. Their evidence suggests that the gap between students from high- and low-socioeconomic backgrounds widened with the introduction of tuition fees in 1998. At the same time, the development of mass higher education means that the role of social networks in gaining quality information about qualifications and the institutions from which to obtain these qualifications have arguably become more important. Adnett and Slack (2007) argue one reason for this is that as participation in tertiary education increases (and the boundaries between groups becomes harder to distinguish), the returns on the qualifications gained may decline, and other signals of capacity increase in importance. What we are witnessing here is the deployment of new strategies employed by individuals, families, and organizations to secure advantage. In the case of networks, the old boys' network may, for example, be of declining importance in some fields as a means of gaining direct access to employment. However, networks are arguably of increased importance in terms of making a number of critical decisions about education, including the choice of where to gain a tertiary qualification. In a similar refrain, the growing importance of networks in gaining competitive advantage is driving the formation of new relationships between groups and individuals across the globe (Saxenian 2002). In relation to this, as resources that were of value in earlier periods have eroded, so too have certain strategies declined in value. The state played an important role in these processes by helping create and police new boundaries.

Thus, although research has improved our understanding of patterns of advantage and disadvantage, another way to understand the relationship between tertiary education and social class is to view changes in the composition of social class groups as an inherent feature of the revolutionary nature of capitalist production. Indeed, the dynamic and changing nature of capitalist production will necessarily mean that the composition of elite or class groups will change in conjunction with economic and social change. In turn, although shifts in production may have changed class-based opportunity structures, this has not meant that society has become classless. Rather, as Erikson and Goldthorpe (1992) note, class formation is undergoing processes of constant flux. This means that we can reasonably expect processes of class reproduction and reformation to be occurring all the time. Thus, we are not seeing the end of class, as predicted by the technocratic-meritocratic perspective, but we are seeing both changes in the way that social relations of production are organized under capitalism and new processes of class formation, which differ according to numerous factors. These include the geographical location in which the processes are taking place, the age of those that experience such processes, and the field in which the processes are played out.

As part of ongoing processes of boundary formation and reformation (and class formation and reformation), then, new rules of advancement and new notions of merit are evolving. Furthermore, they are being developed by the state and by elite groups as they attempt to adjust to new economic conditions. For example, in what can be considered a project in national identity building in New Zealand, the Labour-led government of 2005–2008 has bought together discourses associated with the knowledge economy, creativity, and entrepreneurship to advance New Zealand as a creative economy. In this context, the filmmaker Peter Jackson, who has directed a number of high-profile film productions, has been used as evidence that New Zealand can win in the global competition for advancement. As part of this change, creativity and entrepreneurship are being formulated as an emerging source of advantage for new elites in the global economy. Thus, creative entrepreneurs such as Peter Jackson, along with high fashion designers such as Karen Walker and Trelise Cooper, are now doyenized as the new heroes in New Zealand's attempt to win in the global economy. Developing entrepreneurship depends on individuals developing certain technologies of the self, or techniques that allow individuals to better regulate their bodies, their thoughts, and their conduct. In the context of the creative sector, entrepreneurship includes an ability to network and to create original works. The development of entrepreneurial culture is a social technology that supports and enhances the productive forces found in the creative sector.[17]

At the same time, the state is buttressing this discourse with policies designed to support these new rules. For example, the state now funds training courses linked to the creative sector, has provided the field with tax relief, and financially supports organizations that are advancing the sector's interests, such as the New Zealand Screen Council.

Conclusion

The purpose of this chapter has been to sketch the theoretical terrain on which this book is based. It has shown that although research suggests that inequality continues to be transmitted across generations, we know relatively little about the precise mechanisms through which inequality is transmitted. Although there is some evidence that those who attend elite institutions gain an earnings premium, in general, the picture is mixed with, for example, several factors related to outcomes cutting across the status of institutions. In part, the reasons for such differences are likely to lie in individual background factors, in institutional factors, and in differences in the fields in which graduates work that are not captured by the research methods employed.

In this respect, this chapter has argued that looking at the outcomes of participating in tertiary education across the whole sector has two important limitations. First, looking at the whole sector risks hiding important variations within particular fields. Second, assessing the relationship between qualifications and outcomes says very little about how individuals come to deploy their qualifications. For example, it is likely that in some sectors of the labor market, qualifications make little difference in practice. So, in fields of the labor market where qualifications are a poor proxy of the skills needed to be effective in the workplace, it might be the presence of other forms of capital that makes the critical difference between obtaining—or not—employment in the sector. Similarly, the values ascribed to qualifications in the technocratic-meritocratic perspective may underestimate the importance of network formation as an institutional strategy. Thus, it is likely that behind the value of qualifications are important social processes that affect labor market outcomes.

Although the extent to which social capital (as well as other forms of capital) is of increased significance in various fields remains an empirical question, the demise of Fordist methods of production—in Western nations at least—and the emergence of network production (Castells 1996) suggest that social capital in the form of networks will be of increased importance in the future. Indeed, as argued throughout this book, the development of social capital, particularly in the form of networks, is now a key aim of government policy. Understanding the role of the state in driving network formation as a strategy is useful, as it helps shift the focus away from strategies employed by families (Devine 2004) and to shift their advantage to strategies developed by institutions and used by families. The next chapter further illustrates the case by looking at the role of social networks in the competition for advancement through tertiary education and by considering how organizations and governments might help create geographies of talent by making networks.

CHAPTER 3

Ecologies of Talent

Introduction

The previous chapter argued that exploring processes of social reproduction within fields would improve our understanding of the role tertiary education has in helping to structure competition for advancement. This chapter builds on this argument by exploring the influence of social networks in the allocation of employment and in promoting innovation. This discussion provides a basis to explore knowledge creation and knowledge transfer and the processes of social reproduction that result from this. Granovetter (1995) helped dispel the myth that gaining employment is an open process. The professional, technical, managerial workers he studied relied primarily on their set of personal contacts to find information about employment opportunities, rather than more formal or impersonal routes. He also argued that not everyone pursued jobs via personal contacts. A critical point was that his subjects' social networks structured the possibilities open to them.

It has long been understood that social networks fulfil a number of functions that enhance and restrict employment. However, until Granovetter's (1974) pioneering work, most attention had been focused on blue collar workers, where the available evidence generally supported social network theory. Indeed, throughout the 1980s and 1990s, qualitative and quantitative research consistently showed the key role of networks in allocating labor power, particularly in blue collar and manual employment (see Granovetter [1995], Rosenbaum [2002], and Lin [2001] for reviews of the literature).

In general, the reasons why employers recruit through social networks is well understood. In contrast, the precise functioning of networks and their interaction with other forms of capital (e.g., human capital) across time and space is not. The complexity of the interactions and other issues means that the relationship between networks and advancement remains relatively poorly understood. An additional challenge is that little attention has been devoted to understanding how networks (and, by implication, fields) have been constructed and used by participants to advance their interests.

Recently, attention has been focused on the importance of network creation and maintenance as an institutional strategy designed to enhance competitive advantage. Indeed, network creation and maintenance, as ways to promote knowledge flow, have been linked by researchers and by government to attempts to build high-wage, high-skill economies (Strathdee 2005a). As described in more detail in the Introduction to this book, the basic assumption of this view is that the power of networks, which have been shown to benefit individuals and smaller social groups, can be harnessed to benefit wider social groups. As Anthony Giddens' (1998) *The Third Way* outlines, theories of social capital have been adopted by "third way" administrations (or those that draw on both neoliberal and social democratic approaches to political and economic management; see Chapter 4) in their attempt to build systems of innovation in which work increasingly demands high levels of skill and is highly remunerated. As part of this, attention has been focused on encouraging the development of knowledge economies through network creation. For example, the creation of knowledge transfer networks in the United Kingdom is explicitly designed to promote innovation through knowledge transfer (see Chapter 7). In general, the efficacy of such interventions is supported in the literature, which stresses the importance of developing systems of innovation, including network formation (Florida 2002). An important contribution made by this research has been to highlight how networks can increase competitiveness by, for example, facilitating knowledge transfer and encouraging labor mobility. In this respect, Florida's notion of geographies of talent neatly captures the idea that networks operating locally and internationally provide ways of accessing innovative knowledge (Amin and Thrift 1994; Cooke 2002; Howells 2005).

As noted, the purpose of this chapter is to link network production and employment to innovation. The chapter begins by reviewing the relationship between networks and employment. Following this, it looks more closely at the relationship between innovation and networks. After describing this relationship, the chapter builds on the insights of Hayek (1945) and others to argue that the formation of networks has implications for tertiary education provisions. This discussion provides a basis for the following chapter, which looks at developments in tertiary education policy in New Zealand and England and describes how these contribute to the creation of geographies of talent. Before concluding, the chapter introduces Figure 1, a model of innovation and networking. This model is an attempt to better express the ideas advanced in the book thus far.

Networks and Employment

As suggested in Chapter 1, growth network production is thought to have heralded changes in individual careers, which are seen to be increasingly taking a network form.[1] For example, the concept of portfolio careers attempts to capture contemporary trends in the labor market. It is premised on the idea that instead of working in full-time employment for a single employer, individuals are increasingly working in several part-time jobs (possibly of varying kinds). The emergence of portfolio careers is thought to demand workers with new kinds of skills and dispositions, including organizational skills, career management, and risk tolerance (Jones 2002). As is described more fully later in this chapter, at the same time, changes in production heralded by the emergence of post-Fordism have also arguably increased the significance of networks in production. Thus, networked, or decentralized, systems of production are thought to be an emerging trend. As Porter (1998) suggests, the new post-Fordist firms take a variety of forms. Nevertheless, the maintenance of networks and cooperation between firms and other organizations to leverage advantage is critical. Firms must cooperate with each other, share knowledge, and form linkages for mutual benefit if they are to compete in the new environment.

The emergence of portfolio careers and the apparent increase in the significance of networks in production has implications for the skills, qualities, and attitudes individuals need to hold to gain advancement, and raises questions about how individual competence in the new environment can be signalled. The official view is that individuals need to gain higher levels of qualifications (to meet increased skill demands in the workplace), and that ways of better signalling the skills and attributes of those who hold qualifications are needed (so that employers can be sure new workers have the skills they need to be effective in the workplace).

In terms of the claim that individuals need to be increasingly skilled, or the training gospel, it can be argued that overeducation is forcing individuals to seek out new ways of distinguishing themselves from the mass of graduates now produced by the tertiary sector. In the past, when participation in university study was the preserve of the elite, employers could be reasonably confident not only of the academic ability but also of the social and cultural backgrounds of degree holders. The exclusive nature of tertiary education meant that merely possessing a degree was sufficient to signal competency, and holders were almost guaranteed access into middle-class occupations. However, increases in the number of students entering tertiary education means that this is no longer the case, and employers now need better information about graduates (Brown 2000). To provide this information, new systems of assessing and recognizing learning and capacity have been developed, which create better signals of competency. For example, the adoption of outcomes-based assessment and the related creation of credit-based qualifications are designed to measure and signal ability more accurately (New Zealand Qualifications Authority 1996). However, debate continues over whether or not the new systems deliver on their promises. On one side of the debate, critics argue that such measures have failed to achieve the desired effects and have arguably made matters worse. On the other, proponents argue that although real benefits have been realized, it will take time before the full potential of the reforms is felt (Wolf 2002).

In terms of the claim that better signals of competency are needed, a problem faced by employers recruiting new workers in the contemporary period is that the systems of assessment used in the tertiary sector, and the qualifications produced, have not matched developments in the labor market. For example, some argue that the emerging style labor market has increased the importance of aesthetic labor (Nickson et al. 2003), such as the ability to look or sound "right." One's ability to be successful in the market for aesthetic labor is poorly signalled by educational qualifications.[2] At a more general level, as North (1990) notes, signals of competency (or the rules of the game) are context specific and can reflect macroculture shared norms and practices that widely guide actions and exchange relations. Investing in the right look (Hamermesh and Biddle 1994), being associated with the right people (Kilduff and Krackhardt 1994),[3] gaining a qualification from the right institution (Brown and Hesketh 2004), and adopting attitudes and behaviors that have been endorsed by elites in particular fields (Jones 2002) are all examples of signals of competency. Nevertheless, by providing better-quality information about the capacity of graduates, the attempt to introduce new methods of assessment, such as those founded on competency-based assessment, is an attempt to alleviate difficulties faced by employers making recruitment decisions.

It is important to consider how transactions between employers and job seekers are facilitated in areas of the labor market, where signals of competency are weak and ambiguous or where the cultural skills needed to be effective in the workplace are not easily discerned. A recent judgement by the Scottish Courts illustrates how some employers are attempting to gain better information about the potential of new recruits. This judgement found that Virgin Blue Airlines had discriminated against job seekers applying to be air hostesses who were unsuccessful in their application because they reportedly lacked the "Virgin flair." That is, these applicants were not considered young, beautiful, blonde, and female (Blackley 2005). It appears that recruiters working in Virgin Airline's assessment centers had a clear idea of the aesthetic qualities that make up the Virgin flair and assessed competence by requiring applicants to sing and dance.[4] Bureaucratic systems of organizing the competition for advancement tend to rely on the formation of open systems of accrediting competency, such as educational qualifications. However, as the Virgin Airlines example suggests, educational qualifications do not necessarily effectively signal the skills, knowledge, and qualities employers are looking for. As a consequence, other methods of assessing and signalling competency have been developed to convey the information employers and job seekers are most interested in.

When considering the question of how labor market transactions are facilitated in areas of the labor market where signals of competency are weak, we can usefully look to the work of social network theorists, who have long had an interest in how individuals gain employment and competence. Some social network theorists argue that network recruitment provides a way to help alleviate the challenges faced by employers and potential recruits in the employment process, particularly in settings where information is ambiguous, lacking, or tacit in nature (Granovetter 2004; Lin 2001; Rosenbaum 2002). Indeed, the work of social network theorists has shown that networks transmit reliable information to employers about the potential of new recruits (Granovetter 2004; Lin 2001; Rosenbaum 2002). For example, existing workers know what skills and attributes are required in particular positions, and whether the contacts they have in their social networks also boast the desired qualities. Thus, those who belong to such networks get first access to new employment opportunities. Employers can have faith that workers recruited through such methods will be of suitable quality. A reason for this is that existing workers are unlikely to recommend contacts without the necessary qualities for fear of damaging their own reputation with their employers. In turn, new workers often appreciate the value of the networks that helped them get a job and do not want to risk damaging their own reputation or that of their contacts by performing poorly in the workplace. For such reasons, actors in kin- and community-based networks have been found in some small-scale research to assist new workers into the labor market by speaking for them at their places of work, to have trained them in the skills and behaviors needed to be effective in the workplace, and to police their behaviors once they have been employed (Grieco 1987, 1996). Knowledge about who would be effective in particular occupations was not available through existing bureaucratic assessment systems, or in other codified forms (such as educational qualifications). Rather, it was embedded in the social relationships that resided within labor markets, locally and across time and space. These relationships were created and strengthened through repeated exchange. In other words, network relationships reduce uncertainty by certifying ability, and they also reduce both the costs of making labor market transactions and the risk that poor appointments might be made (Grieco 1987, 1996).[5]

Although Grieco's (1987, 1996) work tends to give emphasis to networks operating locally, it is useful to examine them in global terms. For example, by drawing on case studies of the information technology industry, Saxenian (2002) argues that transnational communities provide a flexible and responsive mechanism for the transfer of skills and know-how. Saxenian (2002) shows how transnational entrepreneurs draw on their technical expertise and networks to transfer skill across international borders in search of competitive advantage sources. In this context, first-generation immigrants, such as the Chinese and Indian engineers of Silicon Valley, have the cultural capital, the language ability, and the technical skills necessary to span social and cultural boundaries between the United States and other nations. Thus, "by becoming transnational entrepreneurs, these immigrants can provide the critical contacts, information, and cultural know-how that link dynamic, but distant, regions in the global economy" (Saxenian 2002, 185). By linking regions, transnational entrepreneurs are able to tap into sources of competitive advantage available in low–labor cost nations, such as China, which are unavailable to people who do not have the necessary cultural or social capital. There is a critical cultural dimension to knowledge transfer because it is easier to build trust within cultural groups that speak the same language than between different cultural groups (Doloreux 2002; McDonald and Vertova 2002). However, it is also important to note that speaking a new language to different social and cultural groups can also be a source of competitive advantage. For example, as described more fully in Chapter 6, by bringing a language of commerce into biological research, chief executive officers in biotechnological companies in New Zealand are identifying new sources of competitive advantage. From the perspective of social capital theorists, such people are boundary spanners, possessing the bridging social capital (Putnam 2000) needed to be able to function effectively in more than one field.

By working across existing fields, boundary spanners also assist in the formation of new fields. For example, having a mix of knowledge of biological process and business skills has allowed individuals to build the field of biotechnology. Pioneers in the field have, for example, developed both high levels of technical competency in biology (and have usually gained high levels of qualifications) and the ability to present sophisticated ideas in ways that appeal to investors (see Chapter 6). In a more academic refrain, Hayek

(1945) has noted that the flow of information in markets is far from perfect, and that opportunities exist for groups and individuals to gain competitive advantage by gaining access to knowledge ahead of their competitors, through exploiting asymmetries of knowledge and information. Actors are positioned in the social infrastructure in different ways to gain insider knowledge and to exploit asymmetries of knowledge by, for example, spanning boundaries. By gaining access to this knowledge, individuals and organizations can take the lead in the competition for advancement.[6]

Similar to Florida (2002), Burt (1997) argues that actors whose networks bridge groups of actors, or structural holes in the social infrastructure, have more network capital or social capital than those whose networks are dense and cohesive. A key reason for this is that actors participating in dense networks circulate information between one another. Moreover, because there are no conduits in the social infrastructure through which new information can flow, much of the information that circulates is already well known by network participants. Much of this is redundant or repetitive and is of little value in terms of gaining a competitive advantage. For example, workers whose networks are limited to areas of the labor market that are experiencing decline are likely to lack information that can help them move to more buoyant areas of the labor market. In contrast, as the work of Saxenian (2002) testifies, actors whose networks provide a bridge between different employment settings are able to gain advantage by exploiting asymmetries of knowledge (and the resources to which knowledge provides access). Moreover, as suggested, they are able to develop competence in different cultural fields and are able to broker information between different groups. By brokering this information, they have power to control how projects develop and to exert control over who has access to valuable information. Those who are able to work with a diverse range of individuals and groups, and whose networks are rich in bridges, are thus able to mobilize more resources and to be more entrepreneurial (Emmison 2003). For example, Benner (2003) argues, on the basis of his research into the information technology industry in Silicon Valley, that functions completed by brokers in matching employers and employees are an emerging business opportunity. In other words, because the matching function is central to competitiveness, it is important that it be done as efficiently as possible. Benner's argument is that business opportunities are present for those who can take on this role for firms. Indeed, as he argues, many human resource firms function by completing this task.

Similarly, the development of employee referral programmes in the United States speaks of the power of network recruitment within the ranks of middle-class occupations (Lachnit 2001), and in lower-level service occupations (Fernandez et al. 2000). Finally, there is evidence that the formation of networks between institutions and employers can improve the transitions of work-bound school leavers (Rosenbaum 2002; Strathdee 2005b).

Networks, Knowledge, and Innovation

To further our understanding of the relationships among networks, employment, and innovation, it is useful to review the developments in production and in tertiary education that are thought to be driving economic competitiveness in the global economy. It is now widely accepted that rapid changes in technology and shifting patterns of consumption are making it difficult for traditional firms to compete in an increasingly global economy (Kanter 1995). Although we need to note variations that exist between fields, an effect of globalization is that companies that remain locked into traditional methods of production, or those that do not seek sources of competitive advantage from across the world, will tend not to survive. When assessing the validity of this argument, it is useful to look to the work of scholars such as Freeman (1997) who have built on the insights of Schumpeter (1934) to theorize the relationships among innovation, technological change, growth, and trade. Freeman's (1997) insights have influenced our conceptualization of the knowledge economy in several important ways.

First, they have highlighted the role played by technological change as a key driver of economic growth in the contemporary period. Of significance is the development of microelectronic technology and related developments in such areas as computer technology, biotechnology, and information technology. The emphasis on technological change, as a driver of innovation, underscores the importance of techno-scientific knowledge in the knowledge economy.

Second, the emergence of new technologies and other sources of competitive advantage have highlighted the need for firms to respond quickly to changes in consumer demand (Porter 1998). Readers will note that such observations do not apply to all sectors of the economy; for example, in instances in which competitiveness is derived through protectionist policies and practices. However, many goods, including those that depend on a labor force with a high level of technical competence and expertise, can now be produced more efficiently in countries with low labor costs, such as China, India, and Mexico. Indeed, shifting production to areas of the globe where the costs are lower is itself an innovation, and one that is helping drive England's and New Zealand's desires to modernize the production of innovative knowledge (Department for Trade and Industry 2003). As Porter (1998) points out, innovation can occur in all sectors of the economy but need not create employment that demands high levels of skill or that pays highly. Indeed, the connection between skill and remuneration is highly context specific, and there exist many examples of work that is highly skilled but poorly remunerated. For example, many forms of craft work, such as welding and carpentry, require high levels of skill, but relative to many professional and white collar occupations, these forms of work remain poorly remunerated. Indeed, as the emergence of increasingly skilled but relatively low-cost labor in places such as China and India demonstrates (Shenkar 2004), competitive advantage can be derived in ways that challenge the official view from New Zealand and the United Kingdom that investment in skill development will lead necessarily to high-paying employment. At the same time, there are many examples of work that require only low levels of skill (or at least that do not require substantial levels of education and training) but are highly paid. At this juncture, it is worth restating a point made in Chapter 1: What matters are the competitive forces at play in specific market contexts and how producers differentiate themselves from their competitors by exploiting, for example, asymmetries of knowledge and other factors of production.

Third, at an organizational level, Porter (1998) has emphasized that the maintenance of networks and cooperation between firms and other organizations, to leverage advantage, is critical to competitiveness. The sources of advantage include the ability of networks to provide access to innovative knowledge and to provide quality information about potential recruits. For individual firms, these networks may operate locally, internationally, or some combination of the two. Again, when considering these comments, it is important to remember that the precise configuration of networks depends on the competitive pressures at play and the systems of innovation present within the field. Whatever the precise mix of network strategies, as Porter (1998) states, competitive advantage derives from differences between firms. No matter what the nature of the competition between firms, a firm's strategy for gaining advantage will be based on the extent to which it can differentiate itself from its competitors. It is possible for firms to differentiate themselves internationally and nationally through clustering. However, within clusters, competitive advantage remains based on the principals identified by Porter (1998).

Fourth, there is no one system of innovation, and factors that give rise to innovation vary from field to field. In this respect, there is a growing acceptance that different (and even conflicting) systems of innovation will operate at different sectoral, transnational, and indeed regional levels (Freeman 1997; Lundvall 1992). Thus, in their own ways, these theorists of innovation systems accept the idea that different sectors (or, to adopt the language of this book, different fields) operate under different technological regimes and have distinctive characteristics (Carlsson et al. 2002).

When considering these issues, it is important to recognize that globalization and the introduction of new technologies have reduced in significance barriers that have traditionally limited the ability of some groups to innovate. For example, in the past, asymmetries in access to financial capital meant that developing countries could not gain the finance they needed to build productivity (Agbetsiafa 1998). Although significant barriers remain, in the contemporary period, the increased availability of new sources of investment capital in developing nations, and new ways of accrediting borrowers, means that companies can more easily access the financial resources they need. Indeed, globalization means that entrepreneurs need no longer be limited to domestic sources of capital. Those who make use of the new sources of finance first will be advantaged. However, over time, relative equality of access to sources of investment capital (and the implied declining returns) means that in the future, competitive advantage will need to be derived from other sources, such as the development of new technologies, increased access to new sources of suitable labor (by, e.g., increasing migration), and increased investing in education and training. In this respect, attempts to open formerly closed labor markets to foreign workers, and the creation of regional, if not global, qualifications systems, are important aspects of an attempt to lift competitive advantage through increasing international labor mobility. However, similar to access to capital, the competitive advantage that is derived from these strategies is subject to diminishing returns, as competing producers adopt similar strategies. What matters in terms of gaining and maintaining competitive advantage is ongoing access to resources that competitors do not have, or protecting yourself from competitive pressures, by ensuring preferential treatment in accessing markets.

Knowledge and Competitive Advantage

When considering sources of competitive advantage, it is useful to distinguish between different kinds of knowledge because it is not all of equal value. The standard approach is to distinguish between two ideal types of knowledge; that is, explicit codified knowledge and tacit knowledge (Simmie 2003). Tacit knowledge is embodied in experts, and its use is neither governed by rules nor codified; it is often implicit or unspoken. Tacit knowledge is necessarily personal, as it is developed through firsthand experience and interaction, and it usually takes two forms. First, tacit knowledge consists of cognitive dimension, which is reflected in beliefs, schemata, or mental models. The cognitive aspect of tacit knowledge is important, as it shapes how individuals see the world. Second, tacit knowledge consists of technical skills, crafts, and know-how, which can be context specific (Nonaka and Konno 1998).

Both forms of tacit knowledge are difficult to transmit in asocial ways; indeed, this is arguably the source of much of their value. Tacit knowledge is critical to many innovations because it involves the generation of new knowledge through discovery; for instance, through acting on intuition. Because tacit knowledge is difficult, if not impossible, to transmit in asocial ways, its transmission depends on processes of socialization, such as those found in apprenticeship training under master craftspeople. Finally, tacit skills are not only embedded in the human capital of individual workers but are also embedded in organizational routines and procedures adopted in the workplace. Thus, teams of workers can build tacit skill that is embedded in the relationships and interdependencies that develop between team members (Grieco 1996).

The social nature of knowledge creation and the importance of this knowledge as a source of competitive advantage can also be seen to relate to broader shifts in how production is organized—principally the demise of Fordist organizations, in which all elements of production were contained in one organizational structure, and the rise of post-Fordist, or network, forms of organization. As noted earlier in this book, it is important to acknowledge that within the debate, differences exist over the term post-Fordism (and its predecessor, Fordism), and that its usefulness in describing reality remains highly questionable (Lloyd and Payne 2002). Nonetheless, the idea is that in the contemporary period, new organizational structures are emerging in which the bureaucracies of what could loosely be described as Fordist education systems (Brown and Lauder 1996) are being replaced by post-Fordist network structures (Castells 1996). As noted earlier, although the new post-Fordist organizations take a variety of forms, the maintenance of networks and cooperation between businesses, the government, and knowledge creators is critical.

The basic premise is that under Fordism, output was based on producing large numbers of commodities and supplying these to mass markets. In this system of production, workers (usually male) performed routine tasks, many of which required little, if any, training. A key innovation associated with Fordist production was the introduction of the principles of scientific management into the workplace. Accordingly, competitive advantage was derived both from workers completing the same small task repeatedly on a production line and through managers' use of workers' experiences and knowledge to drive efficiency gains in the production line.

Workers were given some opportunity to share their knowledge with management and with other workers within their firms. However, there was relatively little cooperation between firms. This meant that the design, development, and production of commodities and other functions were largely completed within large organizations. The high levels of productivity, full employment, and development of the welfare state that arose in concert with the introduction of Fordism meant that the economic and political system was able to deliver improvements in the living conditions of many. However, as competing firms adopted similar strategies, the competitive advantage derived from Fordism declined in Western nations (although its significance in newly industrializing nations such as China and India continues).

The demise of Fordist production in the late 1970s necessitated the adoption of new strategies to achieve competitive advantage. One of these was teamwork, which was designed, in part, to elicit the more active collaboration of the core workers. The knowledge of workers and, critically, their desire to act on this information and to share it with other workers within firms was increasingly seen as a major source of competitive advantage. Thus, in the contemporary period, the development of quality circles, cell production, and related processes of multiskilling can be seen as attempts by management to encourage workers to reveal and share their tacit skill. From the perspective of business, the sharing of knowledge in this way is designed to reduce transaction costs (or the cost of gaining access to workers' knowledge). From a left-wing perspective, the sharing of knowledge can represent the transfer of skill ownership from workers to management and can reduce workers' power to bargain for improved conditions of employment (Jordan and Strathdee 2001).[7] A further way of gaining competitive advantage through accessing the tacit knowledge of workers is to embody it in technologies. For example, Nonaka (1991) describes how a production team at Matsushita Electric Company drew on the tacit knowledge of an acclaimed baker to overcome problems in the design of their mechanical bread-making machine. Through a process of socialization, a member of the production team was able to gain access to tacit knowledge that could overcome limitations in the design of the bread-making machine. Once this knowledge had been obtained, it was articulated to other members of the production team, who incorporated it in the design of a new machine, which went on to establish a new sales record.

As noted above, it is important to consider the relationship between codified knowledge and innovation. In contrast to tacit skill, codified knowledge is explicit and can be transmitted in abstract forms; for instance, in a textbook or a manual. In relation to recruitment, an individual's capacity to understand codified knowledge can be assessed with some confidence through educational assessments, and their competency can be more easily signalled by educational qualifications. As a rule, the more knowledge or experience has been codified, the lower the costs of transporting it to others.

This basic principle underpins recent attempts to diffuse skill or democratize access to knowledge through developing standards of occupational competency as promulgated by the New Zealand Qualifications Authority and the Qualifications and Curriculum Authority in England (see Chapter 4). The idea is that improved systems of codifying knowledge and dispersing it throughout the education and training sector, and the economy in general, will help drive up innovative capacity. It will do this by allowing all to access high-quality, economically relevant education and training. Here, the development of national qualifications frameworks is part of a broader campaign to democratize access to knowledge by breaking down forms of social closure, such as those that reduce competition by limiting access to knowledge (Murphy 1988). It is also an attempt to increase the ability of qualifications to signal expertise and thus improve labor market transactions by reducing employers' reliance on network-based recruitment.

It is also important to note that, although for explanatory purposes it is useful to contrast these different forms of knowledge, in practice, tacit knowledge is often needed to understand formal codified knowledge. For example, although an instruction manual or a circuit diagram may represent a highly codified form of knowledge, tacit knowledge is often required to make sense of them. Moreover, in the cases of tacit and codified knowledge, whether or not those receiving the knowledge can use it depends on the users' proficiency with the codes developed to transmit codified knowledge, as well as the context in which individuals are situated.

Tacit knowledge may be needed to make sense of codified knowledge. However, once knowledge is codified, its contribution to competitive advantage diminishes because it can be transmitted more easily. In this respect, it is possible to equate human capital with tacit knowledge because, in the context of the massive growth in the proportion of people with qualifications, little competitive advantage can be gained by producing increasingly qualified people (i.e., qualifications are subject to declining returns).[8] This point is, of course, open to criticism. One criticism is that competitive advantage will increase as a result of increasing the number of people who hold qualifications. To date, the lack of a critical mass of qualified workers has damaged competitiveness by reducing the attractiveness of nations such as New Zealand and England as places to invest in high-skill, high-wage employment (Brown and Lauder 2001). Another criticism is that an underproduction of graduates increases the labor cost to firms for employing graduates and reduces competitiveness. The overproduction of graduates has the reverse effect.

Despite these and other possible weaknesses, it can be argued that exploiting asymmetries of knowledge provides an important source of advantage: ongoing access to tacit or uncredentialed knowledge is central to competitiveness. Tacit knowledge can be seen as a key asset in securing competitive advantage simply because accessing tacit knowledge and replicating the conditions in which it can be deployed and used is very difficult to achieve and is, consequently, expensive. Thus, unlike many other commodities, tacit knowledge cannot be traded with ease and usually involves transmission from person to person. Of course, this creates tension for firms, which find that although the transmission of tacit knowledge within their organization is necessary, it increases the likelihood that their workers will be poached by competing firms that believe that access to the skills held by insiders will increase their competitiveness.[9] Although there is a paucity of research in the area, it has been argued that recruiting teams of workers is an attractive alternative to merging with other firms. As Groysberg and Abrahams (2006, 133) point out, "lifted-out teams don't need to get acquainted with one another or to establish shared values, mutual accountability, or group norms; their long-standing relationships and trust help them make an impact very quickly." However, Groysberg and Abrahams also highlight the fact that the available evidence suggests that team recruitment has produced mixed outcomes.

A further reason why ongoing access to tacit knowledge is central to competitiveness in so-called knowledge economies and that, in those economies that aspire for such status, innovative ideas and the products they give rise to suffer from codification and standardization as producers seek efficiencies in production. For example, the embodiment of tacit skill in the design of a new bread-making machine was an innovation that led to significant economic gain. However, over time, the value of this innovation eroded as competing producers incorporated similar elements into their own designs. In another example, the combustion engine was a significant technological breakthrough, and its creation reflected significant innovation. However, over time, the production of the combustion engine became sufficiently standardized, and the knowledge necessary to manufacture engines became sufficiently codified, for trained engineers and production workers to produce them in low-cost countries. Ultimately, many products that have emerged from formerly innovative ideas can be produced more efficiently in nations where the costs of production are lower and the profits higher. This process has been facilitated by the development of new information technology that has eased the transfer of codified knowledge across the globe and by related changes in the rules of global competition represented by trade agreements. As Brint (2001) argues, in the future, many current innovations will become subject to standardization and codification. This suggests that if the activities of firms are limited to using formal codified knowledge, they will not be in a position to innovate and will only be able to maintain economic competitiveness if they compete on the same terms as countries in which the costs of labor are much lower.

The Social Nature of Knowledge Creation

It is useful at this stage to say a little more about just how innovative knowledge is created and accessed. In terms of the creation of innovative knowledge, as Hayek (1945) pointed out, knowledge creation is typically a result of collaboration between individuals. The basic premise underpinning this view is that creating the knowledge required to be innovative depends on the presence of interpersonal relations of trust and cooperation (Archibugi and Lundvall 2001; Cohen and Fields 1999; Nahapiet and Ghoshal 2000). Collaborations can include participating in relatively closed learning communities or in open networks, in which information is freely shared.

This basic premise has provided a foundation for numerous perspectives on the sources of innovation and how firms and nations can face up to the pressures created by globalization. For example, Doloreux (2002) advances the idea of a learning economy to describe how innovation is fundamentally social in character and results from individual collaboration. However, it is important to remember that the propensity to collaborate is context specific and that there exists no one kind of work organization that maximizes the likelihood that collaboration will occur. A variation of the view that workplace reform can enhance knowledge transfer, increasing innovation capacity and, hence, competitive advantage, is that the development and transmission of new knowledge is facilitated by the creation of a social infrastructure. As argued more fully in Chapter 4, third way administrations in New Zealand and the United Kingdom are attempting to build social infrastructures conducive to knowledge transfer. Creating structures that disperse knowledge within and between organizations in ways that do not compromise competitiveness is required. The better the social infrastructure, the more effective it will be in creating new knowledge and in transmitting this knowledge to users. This point was well made by Hayek (1945), who believed that progress was derived through knowledge generated locally, as found in markets.[10] In its many variations, the notion that innovation is fundamentally social builds on the idea that all knowledge cannot be concentrated in one individual's mind, and that no single mind can know in advance what knowledge will be of value in the future. For these reasons, the creation of new knowledge is fundamentally a social process involving relations of cooperation and trust (Archibugi and Lundvall 2001; Nahapiet and Ghoshal 2000).

It is important to point out that the predictions of some that production would be increasingly decentralized, or networked, as producers adopted lean methods have yet to come to fruition (Harrison 1994), and that the debate over the effect of firm structure on competitiveness is far from settled (and, given the evolving and unplanned nature of production under capitalism, is unlikely ever to be so). Nevertheless, the theory does speak of important developments in production that are likely to occur in competitive firms of all sizes. Indeed, there are numerous examples, but network formation has been found to aid production in areas such as Silicon Valley (Benner 2003; Casteilla et al. 2000), Oxfordshire and Cambridgeshire in England (Garnsey and Lawton Smith 1998), and in sectors of the labor market, such as the screen production sector in New Zealand (de Bruin and Dupuis 2004), which has grown rapidly in recent years (see Chapter 5).

Tertiary Education, Networks, and Innovation

Along with seeing network formation as a strategy employed by individuals and the firms in which they work to increase their competitiveness, it is useful to conceive of network formation as a strategy used by institutions of tertiary education as they attempt to find new sources of revenue. The notion of networking as an institutional strategy derives from the same idea advanced above, namely, that investment in social relationships pays dividends. In terms of access to scientific/technological knowledge, linkages between innovative firms, entrepreneurs, and knowledge generators are a key aspect of successful clusters, such as Silicon Valley (Casteilla et al. 2000). Casteilla et al. (2000) describe the close, but fluid, relationships among universities, venture capitalists, and firms, which exist in this cluster. Personal links between individuals in these three groups provide a key conduit through which firms gain the latest information about the current research and funding opportunities. For example, through these networks, firms have early access to the latest research reports and have individual meetings with researchers when needed. Moreover, established links between the universities and firms provide a way for job seekers to find employment, both through the recommendation of professors and through affiliated programs. In a similar vein, new recruits are able to increase their chances of gaining employment by using actors in their network to influence recruiters, possibly by asking them to put in a good word or to speak for them.

To speculate further, including innovative firms in decisions about research and development priorities increases the likelihood that the kind of knowledge produced by universities will have commercial value. However, the creation of networks provides a means for firms to have a greater influence on the kinds of human capital produced. Indeed, through working on collective projects, employers are able to gain valuable information about the skills of graduates that are not provided through qualifications. For knowledge generators, the formation of joint ventures offers new sources of revenue through the commercialization of intellectual property, provides a way to improve the employment prospects of graduates, and provides a way to help ensure that the knowledge and research produced has economic relevance.

This perspective arguably helps to theorize the development of reputational capital. It is likely that it is the quality of the interactions that firms have with the graduates of tertiary institutions that underpins the formation of reputation. As argued later in this book, in some fields this will be based on technological and scientific innovations; in others it will be based on social innovations. After all, reputation does not emerge spontaneously; rather, it is a result of interactions between parties and is strengthened by repeated exchange (Grieco 1996).

In this respect, research has confirmed the importance of interactions as a basis for the formation of reputation (Morley and Aynsley 2007). When asked how they judged the quality of a candidate's institution, employers who participated in Morley and Aynsley's (2007) research "revealed that its general reputation and performance of previous selected candidates were the most common criteria used." As one respondent to their survey put it, "as long as we keep taking graduates from [the 20 universities targeted] . . . and those graduates come into the business and perform exceptionally well, there is no reason to change" (Morley and Aynsley 2007, 240). Whether or not reputation reflects employers' satisfaction with the cultural capital graduates from elite universities possess (albeit legitimated by qualifications), as Brown and Hesketh (2004) suggest, remains an empirical question but is likely to vary by field.

Although Morley and Aynsley's (2007) research shows that employers perceived there to be a hierarchy of institutions, when it came to the relative importance they placed on various factors related to employment, as related research found (Morley 2007), the reputation of the university that awarded candidates qualifications was well down in the employers' list of priorities when selecting new recruits. Indeed, reputation of the awarding institution (scoring 90 out of a possible total score of 200) ranked below a range of soft skills, such as interpersonal/team skills (173), and behind degree classification (146) and subject knowledge (122).[11]

In relation to this latter point, knowledge generators are able to build their reputations by building relationships with employers in innovative firms. In turn, it is possible to develop the argument that the extent to which they are effective in transmitting innovative knowledge to students, those knowledge generators with the most innovative staff, with the biggest research and development budgets, and with institutional structures that encourage the commercialization of intellectual property are likely to be best placed to build networks (and increase their reputations) with innovative firms and also to be capable of helping graduates obtain employment in innovative firms. The formation of relationships between innovative firms and knowledge genera-tors, particularly universities, is underpinned by economic and institutional factors that have long been present in areas of innovation, such as Silicon Valley (Cohen and Fields 1999). One economic factor is that the pace of technological change and development is contributing to a blurring of the boundaries between basic and applied research and is increasing the signifi-cance of basic research in the production of new commodities. This can be seen in the development of joint research and development projects between innovative firms, universities, and financiers (Lam 2002). As Florida's (2002) work suggests, geographies of talent are required to facilitate this develop-ment. Another economic factor is that innovative knowledge, or knowledge that leads to competitive advantage, is unavailable to competitors. Thus, innovation in knowledge economies, because of its social character, is likely to come from the creation of new networks or the creation of linkages between research and development organizations, particularly universities, which are directly linked to innovative firms and to sources of financial investment. These linkages provide a critical method of knowledge creation and, criti-cally, the commercialization of intellectual property.

Techno-scientific knowledge (and arguably other forms of innovative knowledge), investment, networks, and innovation all feed into a concep-tualization of the knowledge economy as national systems of innovation. In addition, it is a nation's public and private sector institutions and their activi-ties and interactions that both create and diffuse new technologies (Freeman 1997). It is this kind of view that helps to justify the role of the government as a broker of networks. For example, New Labour's role in matching ideas with wealth through knowledge transfer partnerships (see Chapter 7) is seen to create an "explosive effect," which "sparks a chain reaction to jobs, wealth and higher productivity" (Johnson 2006, 1).

Given the relationship between social capital and innovation, it is not sur-prising that the Organisation for Economic Cooperation and Development sees network formation as an integral aspect of national systems of innova-tion. The Organisation for Economic Cooperation and Development states: "the configuration of national innovation systems, which consists of the flows and relationships among industry, government and academia in the devel-opment of science and technology, is an important economic determinant" (Organisation for Economic Cooperation and Development 1996, 4).

Model of Fields and Capitals

Despite the attractions of the idea that innovation (and hence competitive advantage) is derived through relationships between people, it is important to reiterate that innovation is derived in different ways in different fields. The following model is an attempt to show more clearly some of the possible configurations of know who and know what while incorporating a measure of quantity of contacts that are needed for advancement within discrete fields. Readers interrogating the model will no doubt identify problems; nevertheless, the three-dimensional aspect of the figure is presented in an attempt to capture some of the complexity of the interactions that occur in the competition for advancement in different fields.

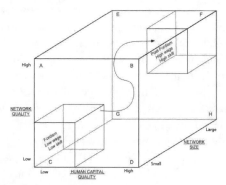

Figure 3.1 Model of Fields and Capitals

In this model, network quality refers to the quality of the resources contacts and individuals need to access through networks to succeed in a field. Human capital quality refers to the human capital, broadly conceived, that individuals need to possess to advance in a field. Finally, network size is intended to indicate the size of the networks needed by individual actors for advancement within a field.

The model attempts to capture the fact that in some fields, only low levels of human capital and poor-quality networks, which are small in size, are needed for advancement. For example, many fields make few, if any, demands that new entry-level recruits hold high levels of human capital. Nor does gaining entry to such fields require new recruits to participate in large networks that contain high-quality contacts. As indicated in the figure, examples of such fields include many low-wage/low-skill forms of employment, such as sales positions in established retail settings or supermarket checkout operators. These would be positioned at, or close to, point C in the above model. A different set of resources is needed for other sales-type positions, such as those in real estate, where large networks that provide access to high-quality resources and are required for success. Such fields could be positioned around point E in the above model (another view would be that selling real estate requires high levels of human capital and be better positioned toward point F).

Other fields require high levels of human capital, yet the recruitment methods present are based on advertisement and interview. It is true that in such fields, network quality and density can be helpful in alerting job seekers to vacancies and can improve an individual's chances of success (through, e.g., providing them with insider status and a supportive voice within companies [Granovetter 1995]). However, in theory, neither network size nor network quality is as important as human capital (as expressed by formal qualifications for example). Examples of such fields include those where bureaucratic recruitment methods are employed, such as teaching and nursing. Such fields would be positioned at, or around, point D in the above model.

At point F are labor market fields in which all three attributes are important. As we see later in the book, biotechnology comes close to this position. This is because in this field, high levels of human capital, quality contacts, and large networks are all required for entry and advancement. At point B are those forms of work that demand high levels of human capital and networks that are of high quality. Network size is relatively unimportant. It is more difficult to provide examples of occupations in which this mix of resources is required, but a case can be mounted that changes in technology and other competitive pressures, which increased the skills and training needed to be successful, mean that entry into family businesses has moved from point A to point B. The arrow from the box at point C to the box at point F is intended to show how modern states are attempting to transform the labor market away from providing predominantly low-wage/low-skill and individualized forms of work and toward providing high-wage/high-skill and networked forms of work.

Two further aspects are worth noting: first, the model allows for fields to evolve. For example, in the past, father-to-son patterns of inheritance in areas like farming would have seen farming placed at point A in the model. However, changes in agricultural technology mean that it is increasingly important that farmers have high levels of expertise and training. At the same time, in an attempt to remain competitive, smaller, uneconomic farms have been incorporated into larger enterprises. These and other pressures have seen the rules of advancement in farming shift from point A to now include a greater role for human capital and an increased role for network size (i.e., toward point F). Second, the model allows for circumstances of time and place. Differences in local conditions might decrease the importance of knowing what in some fields relative to the level required in other contexts. For example, gaining a professorial position in a prestigious university arguably requires greater evidence of proficiency in knowing what than that required in a less prestigious university.

Conclusion

The central purpose of this chapter has been to draw links between social networks, innovation, and competitive advantage. It has drawn on existing arguments to sketch a broad overview of the relationships among network creation, innovation, and the competition for advancement.

It has shown that a key reason network formation is high on the agenda of governments in New Zealand and the United Kingdom is that innovative knowledge is resistant to codification. Indeed, the construction of networks suggests that third way governments in these nations have lost faith in the ability of bureaucratic and egalitarian systems of competition for advancement. Network creation is promoted by some on the basis that it will distribute information widely or democratize access to knowledge (Szreter 1998).

Although there is mounting evidence that networks among providers of tertiary education, employers, and sources of investment are important in areas of the economy that depend on scientific/technological knowledge, there are good grounds to believe that such networks are also important in other areas of the economy. In recent years, the number of tertiary education providers producing graduates for the creative sector has increased dramatically. However, very little is known about the relationship between providers of tertiary education and employment in the sector. Finally, the Model of Fields and Capitals was introduced as a way of illustrating how the rules of advancement within fields vary.

Chapter 4 describes how third way administrations are attempting to create geographies of talent by bringing providers of tertiary education, particularly universities, into a closer association with industry and by considering the effect of these developments on social exclusion. For example, although Massey et al. (1992) argue that the ability of such linkages to drive up innovative capacity is likely to be overstated. The New Labour government in England hopes to drive up the production and dissemination of innovative knowledge by building knowledge transfer partnerships between different organizations.

CHAPTER 4

The Creation of Contrasting Geographies of Talent in England and New Zealand[1]

Introduction

As noted in Chapter 1 of this book, one weakness in current research into the impact of tertiary education is that it tends to assume that similar rules of advancement operate throughout the labor market. Although research demonstrates a relationship between educational qualifications and employment outcomes, such research does not sufficiently acknowledge that the value of qualifications differs in different fields. For example, although educational qualifications are a prerequisite to advance in some areas of the labor market, they are not as commonly required in others, such as sales and other service sector occupations. Policy makers in New Zealand and the United Kingdom also tend to assume that similar rules of advancement operate in all areas of the labor market. For this reason, they have devoted considerable energy to encouraging all people to engage in education and to achieve educational qualifications, irrespective of the rules of advancement in operation in the fields where graduates hope to work. The introduction of national qualifications frameworks in New Zealand, England, Scotland, and Wales, for instance, was partially predicated on a desire to create systems of qualifications that would drive up innovation by motivating all to succeed in education and training and to demonstrate this by gaining qualifications. Nevertheless, as shown in previous chapters, in some fields, the qualifications gained by individuals have struggled to attract much interest from employers. This suggests that labor market relationships are facilitated in these fields in other ways.

In addition to their role in promoting achievement, frameworks are also an important method of increasing the supply and availability of skills and of diffusing these throughout the economy. According to "third way" governments, skill development and skill diffusion are needed so that economies based on a high level of skill can be established. In turn, the creation of high-wage high-skill economies helps in the struggle to create socially inclusive societies. For example, in England, the recent Skills White Paper (H.M. Government 2005) links skill development and the creation of a new national qualifications framework (known as the Framework of Achievement) to increased innovation and reduced social exclusion. Similarly, in New Zealand, the National Qualifications Framework has assumed a key role in the Growing an Innovative New Zealand Framework (which was subsequently incorporated into the Labour-led Coalition's Economic Transformation agenda [Cullen 2007]). At the heart of the Growing an Innovative New Zealand Framework is a desire to drive up social inclusion through promoting growth in high-wage and high-skill work. Achieving both relies on significant investment in skill development (Office of the Prime Minister 2002).

This chapter argues that skill development, skill diffusion, and the creation of high-performance workplaces through frameworks remain central to attempts in England and New Zealand to foster innovation. However, recent developments in policy suggest that third way governments in these nations are also pursuing network formation as a way of promoting innovation. For example, in England, the Department of Trade and Industry now sees network formation as an important part of its work and is introducing a number of specific initiatives, including knowledge transfer networks, to promote network formation (see Chapter 7).

Although it needs to be noted that the account offered below is highly stylized and abstract, this chapter argues that the current skill strategies of New Labour (in England) and the Labour-led Coalition (in New Zealand) are helping to construct contrasting geographies of talent. Accordingly, the purpose of this chapter is to show how the state affects the rules of advancement within fields. This chapter argues that frameworks are central to the construction of open geographies of talent in which individuals compete on an equal basis for positions in the labor market. By increasing both access to training and the quality of information available to users of education and training, frameworks are intended (at least officially) to contribute to openness across all fields. For example, they are seen to break down forms of social closure that have, to date, limited the ability of outsiders to make good decisions about education and training. In this respect, frameworks are seen to increase the ability of employers to adopt a job-competency approach in which the skills and attributes needed to succeed in a given position are clearly stated. In turn, this increases the ability of job seekers to know what competencies they need to hold if they are to win positions in the labor market. Employers also benefit from an increase in the reliability and trustworthiness of information that is perceived to be provided by frameworks about the potential of new recruits.

Frameworks can also be linked to greater democracy. One way they are thought to contribute to increased democracy is by providing a way to codify knowledge, so that it can be disseminated widely (Szreter 1998).

In addition to their contribution to increased openness and democracy, it is probable that the skills strategies employed by New Labour in the United Kingdom and the Labour-led Coalition in New Zealand are helping to construct closed (or embedded) education and training markets through creating process-based trust, as exemplified in social networks. The creation of process-based trust is driven by the growing realization that, in some fields, competitive advantage is more likely to result from forming networks than from simply increasing the proportion of individuals who hold institutional-based forms of trust, particularly in the form of educational qualifications.[2] However, although network formation makes economic sense, it sits uncomfortably with New Labour's and the Labour-led Coalition's discourse of social inclusion.

The chapter begins by exploring the relationship between the general aims and methods of Labour governments in New Zealand and England, and their approach to education and training markets. This provides the basis for subsequent sections, which in different ways assess the relationships among markets, networks, and innovation.

The Third Way

The aims and methods of third way administrations in New Zealand and England have been the subject of considerable debate (e.g., Thrupp 2001); there is no need to repeat this in detail here. However, in terms of the arguments advanced later in this chapter, it is important to briefly outline third way approaches to social and economic management as they pertain to education and training.

At a basic level, third way administrations believe that the judicious use of market forces can contribute greatly to the state's social and economic objectives. To take the English example, in contrast to Old Labour's wholesale rejection, and the Conservative's wholesale and uncritical acceptance of free markets, New Labour believes that the ability of market forces to do good should not be dismissed out of hand. It also believes that carefully crafted market strategies can help achieve social goals. From this perspective, it follows that market strategies in themselves do not create social exclusion; rather, it is the way in which neoliberal administrations deploy such strategies that creates the problem. In the area of education and training, for example, the market policies of previous neoliberal administrations led to an underinvestment in skill development. By limiting the state's role to simple support of free markets, neoliberalism actually reduced competitive advantage by failing to support skill development in instances where market incentives to invest in skill development are weak. The Department of Trade and Industry expresses the point as follows:

> We reject the interventionist command and control industrial policy of the past. Equally, we reject the idea that Britain does not need a strong voice for business at the heart of government. Instead, we need a new [Department of Trade and Industry], equipped to deliver in the areas most crucial to our success in the global knowledge economy. A Department of Trade and Industry with a mission to promote world-class science and technology, support British business success and ensure fair and flexible markets. A DTI committed to intervene only where there is clear market failure, and able to create the institutions and partnerships—regionally, nationally and within Europe—that the modern economy demands. (Department of Trade and Industry 2004, 3)

A further problem linked to the uncritical use of market forces by neoliberal administrations is that they have cemented the position of already advantaged groups. It is not that markets inherently benefit elites; rather, the way neoliberal policies were structured and deployed by Conservative administrations led to elites gaining most from the new environment. In response, New Labour claims it is attempting to do more to undermine the status of elites. For example, as noted in earlier chapters, conservative policy and practice are perceived to have barred segments of the population from participating in tertiary education. A range of measures, including providing additional support for those from disadvantaged backgrounds to stay on in education and to achieve highly, has been introduced to rectify this situation. Moreover, procedures governing admissions to higher education have been reviewed to make the system more transparent and, it is hoped, fair (Admissions to Higher Education Review 2005). New Labour is also requiring universities to do more to attract students from disadvantaged backgrounds. For example, an Office for Fair Access has been established to promote and safeguard fair access to higher education for underrepresented groups. This move was deemed to be particularly important given the introduction of

variable tuition fees in 2006. All publicly funded providers of higher education in England who decide to charge tuition fees above the standard level
are required to submit access agreements to demonstrate how the universities
will ensure fair access.

New Labour also argues that neoliberal policies, as introduced by the
Conservatives, were primarily concerned with increasing private ownership
of the economy, not increasing competition (Brown 2003). Free markets,
as opposed to privatized markets of the kind created by Thatcherite neoliberalism, are perceived to be free of class, race, or other such social and economic divisions. To date, the superior ability of elites to access and process
market information, and to access the financial resources needed to put this
knowledge to use, has enabled them to enjoy unfair advantage in the market
(Szreter 1998). Indeed, elites are more embedded in the social infrastructures
that perpetuate advantage. For example, they are more embedded in social
networks, which assist them in the competition for advancement through
education (Ball and Vincent 1998).[3] This is not to say that nonelites do not
participate in networks; rather, it is to say that the networks in which elites
participate provide greater access to assets and resources within fields. It is
also important to remember that elites also exert greater influence over how
fields develop. In other words, elites are best placed to control the formation
of fields (e.g., by determining where boundaries lie) and are therefore better
placed than nonelites to take advantage of any new rules that result.

One result of elites' superior access to, and control over, networks and
the way markets have been created by neoliberal administrations is that the
benefits of economic growth in England have been captured by relatively few
individuals. Another is that the potential contribution that could be made
by nonelites, or those who do not participate in elite networks, remains sup-
pressed, limiting the overall efficiency of the economy. For example, in England, New Labour has been concerned that elite universities have tended not
to recruit students from working-class backgrounds. In turn, young people
from working-class backgrounds have been reluctant to consider applying to
elite institutions, in part because they perceive them to be alien (Archer and
Hutchings 2000). Consistent with the technocratic-meritocratic perspective that was described in previous chapters, this situation has weakened the
economy by reducing the availability of talent.

Reversing this situation necessitates the destruction of such cultures and of
the closed social infrastructures that support them. Policy makers in England
are attempting to build new inclusive cultures and create open social infrastructures by, for example, providing better careers advice via the Connexions
Service. By providing more open and accessible information and advice, the

Connexions Service can be seen as an attempt to repair deficits in the social infrastructures in which working-class students participate. Put another way, the service is designed to compensate for forms of capital that no longer support the state's goals (Strathdee 2005a). Another way the state is attempting to compensate for a lack of capitals is through encouraging universities to take a more holistic approach to recruitment by, for instance, looking beyond an applicant's examination results to include consideration of their social class backgrounds (Admissions to Higher Education Review 2005)[4].

In New Zealand, since the late 1990s, Labour-led governments have also set about transforming the New Zealand economy and its education system to increase innovative capacity through growing more talent. Similar to New Labour, the Labour-led Coalition believes that the introduction of greater competition into tertiary education by previous neoliberal administrations has had a deleterious effect on the economy and social inclusion. As they state, the "competitive model in tertiary education has led to unsatisfactory outcomes in terms of both the quality and the appropriateness of the skills produced" (Office of the Prime Minister 2002, 5). Thus, similar to New Labour in England, in settings where the labor market signals are weak, government intervention is required to facilitate transitions and to improve the effect of the government's investment in tertiary education in ways not possible under the choice model of education (Lauder and Mehralizadeh 2001). To achieve this, the government is shifting money from programs and courses it perceives to be of low value and of poor quality toward courses deemed to be of high value and high quality. A key technique employed by the New Zealand Government to achieve this shift is through funding universities according to profiles. At a basic level, a profile is a tertiary education provider's outline of what it will contribute to the government's priorities, as set out in the Tertiary Education Strategy and Statement of Tertiary Education Priorities. Profiles are subject to negotiation and agreement with the Tertiary Education Commission. The Tertiary Education Commission attempts to identify areas of duplication (and gaps) in delivery with other tertiary education providers. Once a provider's profile has been finalized, it forms the basis for a substantial proportion of its public funding. Funding is typically provided for a three-year period,[5] after which the profiles are renegotiated. Profiles also provide a way of monitoring the performance of providers. Assessment of a Tertiary Education Organization's performance will be against a range of quality measures, including the number of students who graduate and their employment outcomes. Finally, in contrast to the previous system, in which providers of tertiary education were rewarded on the basis of the number of students they enrolled, in the new environment numbers will be capped. In sum, through its Economic Transformation Agenda, the Labour-led Coalition aims to grow more talent through building a quality education system, investing more in industry training, and strengthening pathways from school into tertiary education and onto the workforce.

However, although both New Labour and the Labour-led Coalition have advanced a discourse of social inclusion through education, some of their reforms have arguably made access to higher education more difficult and have increased competition. For example, both administrations have been unable to prevent increases in the tuition fees charged by providers of tertiary education. Nevertheless, the Labour-led Coalition has attempted to reduce the cost of participating in tertiary education by providing students with interest-free loans and by limiting the extent to which institutions can increase their fees. As is the case in the United Kingdom, the idea is that fee increases should not increase social exclusion. At the same time, the Labour-led Coalition has forged ahead with its desire to promote diversity of provision and to concentrate research capacity in relatively few institutions. Funding providers of tertiary education according to agreed teaching profiles is an important strategy geared to achieving this. Another is the introduction of the Performance-Based Research Fund,[6] which is loosely based on England's Research Assessment Exercise.

In the English case, the Research Assessment Exercise has arguably exacerbated differences between older research universities, which have relatively low staff-to-student teaching ratios, and those that do not (Leathwood 2004). In the New Zealand case, the introduction of the Performance-Based Research Fund system has also proven to be an effective method of increasing diversity of provision by separating research institutions from teaching institutions. The introduction of profiles will exacerbate these differences as the government seeks to reduce competition in the tertiary education sector and to find ways to encourage providers of tertiary education to produce graduates with the knowledge and the skills it perceives to be of value (rather than leave it to the market to determine what is valuable knowledge).[7]

Both the Research Assessment Exercise and the Performance-Based Research Fund systems represent somewhat of a reversal of an earlier approach, in which uniformity of provision was a key means of realizing equality of opportunity (although it remains an unachieved goal). In the contemporary period, third way administrations have continued to promote a discourse of social inclusion, though they now do it through discourses of diversity of provision and knowledge transfer.

As noted earlier in the book, it is possible to identify two key, and contrasting, strategies for promoting skill development and innovation. The first is to create national qualifications frameworks, which are designed to increase knowledge transfer and promote skill development through increasing access to knowledge. The second is to create social networks, which are designed to help create knowledge ecologies. The following sections review these contrasting strategies for promoting innovation in greater detail and relate them to social inclusion in greater detail.

Qualifications Frameworks in Education and Training

Although it is recognized by both the state and researchers that creating competitive advantage requires more than simply developing human capital (Brown and Lauder 2001), investing in skill development is a major focus of current efforts by New Labour and the Labour-led Coalition to build high-wage/high-skill labor markets and create socially inclusive societies. The introduction of qualifications frameworks has been a key intervention designed to improve both the production of skills needed to innovate and the successful dissipation of these skills throughout the economy. These frameworks have been introduced at varying levels of the education and training systems in New Zealand, England, and many other nations (e.g., those in the European Union; see following for more details). As the framework qualifications of nations are recognized by other nations, and as systems of credit transfer develop, the likelihood of an emerging global qualifications system increases.

The qualifications framework for England, Wales, and Northern Ireland was reviewed in 2004. The original framework comprised eight levels. Each level corresponded to expected standards of achievement, ranging from the secondary school level through to the postgraduate level. A key aim of the original framework was to bring together all education and training throughout the whole sector into the one, unified, system. It was also intended that qualifications would be broken down into smaller components, so that they could be understood by users. Although this was the intention, in practice the system was far from unified, with traditional qualifications such as A-levels sitting outside the framework. It was also hampered by a tendency to offer only whole qualifications rather than smaller components of learning (Qualifications and Curriculum Authority 2005). The review of the framework expanded on the idea that all qualifications throughout the education and training sectors should be recognized and rewarded in the same way. Following the review, the existing framework was expanded into the Achievement Framework, which is also underpinned by a unit- and credit-based system.

The creation of the Achievement Framework was followed by the White Paper on 14–19 Education and Skills (Department for Education and Skills 2005). The White Paper emphasized a need to make learning more responsive to employer and learner needs and called for a simple and flexible educational currency for learners. A key proposal of the White Paper was that a single qualifications framework for both academic and vocational learning be developed, and that existing qualifications be absorbed into the new system. The review also advanced a modularization agenda. Modularization is designed to break whole courses down into smaller parts, so that achievement can be more clearly reported (Department for Education and Skills 2005). However, New Labour rejected the idea that a single framework be developed. This means that A-levels and General Certificates in Secondary Education remain separate qualifications. However, elements of the General Certificates in Secondary Education, A-levels, and diplomas will be shared, so that progression through various qualifications and levels is facilitated. New Labour is also developing a similar framework for higher education qualifications. Again the aim is to specify more clearly the skills and qualities of graduates from English universities and colleges. There are currently four levels, ranging from certificates through to doctorates. Finally, New Labour has endorsed the Bologna Declaration of June 19, 1999, which led to the development of the European Qualifications Framework and the European Credit Transfer System for vocational education and training. The Bologna process involves agreement about the size of qualifications and the formation of a system of academic grades, which are easy to read and to compare, and which improve international transparency to facilitate academic and professional recognition of the qualifications produced in different nations. By providing detailed information on the curricula and their relevance toward degrees, the development of a system of accumulation and transfer of learning credits is central to achieving this transparency. Increasing transparency is also thought to help establish European markets in education and training and facilitate the migration of skilled labor throughout Europe. By facilitating labor mobility, it is also hoped the measures will increase knowledge transfer and drive up competitive advantage across Europe. The idea is that, to date, the competitiveness of Europe as a whole (and its member countries individually) in the global economy has been hampered by a lack of labor mobility. Increasing the ability of skill workers to migrate throughout Europe is seen to assist in competitiveness by allowing workers to move to areas where there exists strong demand for their skills. In the past, forms of closure hampered the ability of European nations to access the skilled labor they needed to be competitive. As the European Commission put it, "[c]ompetitiveness means putting skilled people where the jobs are. So Europe needs movers!" (European Commission 2003, 4). Participation in the Bologna process is voluntary; nevertheless, over 40 countries have endorsed the process.

A similar situation to that found in England exists in New Zealand, where the New Zealand Qualifications Authority established a national qualifications framework in the mid-1990s. One of the New Zealand Qualifications Authority's functions is "[t]o develop a Framework for national qualifications in secondary schools and post-school education and training" (Government of New Zealand 1995, 242). Central to the framework is the development of unit standards, or discrete learning outcomes, that form the building blocks for higher qualifications. More recently, achievement standards have been developed and placed at different levels on the framework to recognize learning in academic areas of the school curriculum. Unit standards vary in size depending on the amount of work needed to complete them. There are ten levels of learning on the framework, ranging from National Certificates (which are awarded at all levels but normally found at levels 1–4), to National Diplomas (which are awarded at level 5 and upward), through to Doctorates (which are awarded at level 10). The New Zealand Qualifications Authority also developed and manages a system of accreditation designed to ensure that providers of education meet basic standards of service provision. Before the delivery of training that is linked to the framework, and before they are eligible to receive government funds, both private and public sector institutions must be accredited by the New Zealand Qualifications Authority or an organization acting on behalf of the New Zealand Qualifications Authority, such as the New Zealand Vice-Chancellors' Committee or the New Zealand Teachers Council. As is the case in Europe, an important justification for the framework is that it will ensure the future viability of qualifications by providing them with currency on an international stage. It is argued that the development of qualifications frameworks in a number of other Western nations such as Australia, England, Scotland, and the United States is leading to a global qualifications system that will eventually render redundant those qualifications that are not part of this system.

A key innovation of frameworks is the way they allow users of qualifications—particularly employers—to exert greater influence over skill development. In the case of New Zealand, this is best seen in the creation of the system of industry training organizations. A key role of these organizations is to develop new qualifications to support the sectors they represent. So, for example, the New Zealand Equine Industry Training Organisation has developed the National Certificate in Equine (Racing Stable Management) for people wishing to work within the equine and related industries. Qualifications include a collection of units of learning, or unit standards (in the case of vocational learning) and achievement standards (in the case of nonvocational learning). Whole qualifications and the standards that are combined in various ways to make whole qualifications are registered at various levels on the National Qualifications Framework.

Much of the early motivation for developing the New Zealand Framework was that all assessment in the education sector should be uniform. Although it was intended that universities would participate fully in the reform, problems convincing them of the merits of assessment linked to the framework meant that although university-level qualifications are registered on the framework, universities have retained the ability to accredit providers and use assessment systems they deem appropriate. Outside of the universities, providers of state-funded tertiary education have all adopted the framework and related systems of assessment and accreditation (as they must do to gain government funding). It is also noteworthy that universities no longer have the sole right to award degrees, so some providers of degrees, including providers of doctoral degrees, are currently accredited by the New Zealand Qualifications Authority, whereas others are accredited by the New Zealand Vice Chancellors' Committee (on behalf of the New Zealand Qualifications Authority). There are, of course, other forms of accreditation that universities must achieve. For example, universities are significant providers of preservice teacher education and must meet standards of provision approved by the New Zealand Qualifications Authority (which devolves responsibility for accrediting providers and their courses to the New Zealand Teachers Council).

Frameworks, Innovation, and Social Inclusion

Proponents of qualifications frameworks argue they increase innovative capacity and promote social inclusion in a variety of ways.

First, frameworks aid knowledge diffusion by creating national repositories of knowledge and by providing an infrastructure through which knowledge can be distributed. In turn, they reduce social exclusion by increasing individuals' access to knowledge, irrespective of their geographical location, employment situation, or social class. In New Zealand's case, the development of the Economic Transformation Agenda[8] will have an important effect on what providers of tertiary education offer by way of accredited education and training (Cullen 2007). Nevertheless, in general, qualifications frameworks have increased access to codified knowledge. One way they have achieved this is by broadening the range of institutions that can offer training. Thus, they have reduced the ability of individual providers to monopolize provision. They have also arguably increased access to knowledge by helping increase access to knowledge created in various areas of innovation. Indeed, through frameworks, knowledge is transferred through various methods and systems from those who have created it, and it is then converted into codified forms, including standards of achievement, unit standards, and other learning outcomes. As a consequence of this codification, all accredited providers of education and training are free to offer this knowledge to learners.

In this respect, as the Department of Trade and Industry's Innovation Report points out, government has an important role in establishing measurement standards because firms can not exclude competitors from accessing positional knowledge (Department of Trade and Industry 2003). In a similar way, frameworks have a critical role to play in promoting innovation through diffusing knowledge to all individuals in society and in reducing forms of social closure that have allowed groups and individuals to increase the value of knowledge by limiting access to it. In turn, this creates new pathways for learners into, and throughout, education and training, as well as to the labor market.

Second, by facilitating new ways of structuring education and assessing learning, frameworks increase the contribution made by education systems to egalitarianism. In turn, this helps drive the potential to innovate by allowing all learners, irrespective of their cultural backgrounds or current social and economic circumstances, to develop further and to demonstrate their talent. As noted earlier in this chapter, the idea is that competitive advantage depends on accessing the best recruits. To date, selective assessment systems, such as those based on norm-referenced assessment, have primarily benefited elites who are more likely to possess the resources needed to succeed. Indeed, in so-called selective systems of assessment, success was seen to be largely dependent on possessing cultural and social skills, which elite groups tend to possess. Moreover, in selective systems of assessment, achievement tended to be reported as a percentage, or as a simple grade. This further disadvantaged students from nondominant groups because a lack of quality information about their potential encouraged employers to rely on other ways of assessing competency. For example, it encouraged recruitment through closed methods (such as social networks), rather than through open methods (such as advertising and interviewing). By introducing greater openness in the assessment of competency, by increasing the ability of learners to work at their own pace, and by signaling capacity in clearer ways, frameworks are seen to allow all participants the potential for high achievement and progression to all levels of the labor market, irrespective of their background. Here, third way governments are drawing on the longstanding belief that education and training systems, and the qualifications they produce, can compensate for background social factors associated with inequality. Indeed, frameworks are designed to erase geography, so that no matter where or how an individual learns, quality and consistency are maintained. Expressed in a different way, one aim is that the qualifications produced by frameworks will function as forms of institutional-based trust. In earlier times, trust and reputation were based on repeated informal interpersonal exchanges (Grieco 1987). Where reputation became associated with a particular ethnic or family grouping, membership of these various groupings was used as a proxy for other skills and attributes. An outcome of this was the creation of social closure, which contributed to the reproduction of existing social class groups. Moreover, those more closely embedded in the social infrastructure have been better placed to access resources that enhance their transitions into the labor market. These have included information about the skills and qualities needed for employment as well as information about how best to signal that one has these. The development of process-based trust was possible in traditional societies, as labor was not highly mobile, and the skills necessary for production could be obtained through informal methods of skill transference. However, according to the theory, changes in the labor market, and particularly those in relation to skill changes and labor mobility, as well as growing concern with social justice, have rendered process-based forms of trust obsolete and inefficient (Zucker 1986).

In theory, the development of educational qualifications as a form of institutional-based trust represents a victory for egalitarian interests, as all individuals can participate in the labor market to their full potential. In other words, qualifications have been created to replace those functions formerly completed by process-based trust. According to proponents, then, a key advance of framework qualifications over previous systems of delivering education is that they erode the value of resources embedded in the social infrastructure and which are subject to social closure, such as social networks (Strathdee 2005b). In turn, the creation of better forms of institutional-based trust will replace forms of social capital that have allowed individuals of high-socioeconomic status to gain access to positions beyond their "true" capacity. For example, the development of institutional-based trust provides a way to challenge forms of advantage that result from attendance at particular kinds of institutions, rather than from advantage that is derived through actual merit. Indeed, there is no reason why the current situation should continue, whereby graduates from elite universities earn more than graduates with similar qualifications from other universities (Leathwood 2004).

Third, frameworks increase the correspondence between education and training systems and the labor market by making meeting the needs of learners and employers a priority. Prioritizing the needs of learners and employers means that education and training systems are better placed to produce the skills needed to promote innovation. Until recently, the use of conservative assessment technologies (which were preferred by some providers of tertiary education and their teachers) has meant that qualifications did not effectively signal to employers and other interested parties what skills and abilities individuals actually held. One result of this was that that qualifications were used as proxies for other skills that specific groups, such as employers, were more interested in. This contributed to credential inflation, because in the absence of useful information, employers sought out ever-increasing levels of qualification (Collins 1979), which discriminated against students from low-socioeconomic backgrounds, because they are less able than their peers from high-socioeconomic backgrounds to remain in education and training for extended periods.

Fourth, by opening the provision of learning to a wider range of education and training institutions, frameworks provide a way to expand provision and to introduce greater competition into the sector. In turn, this helps create new pathways both into and throughout the education and training sector. For example, frameworks have facilitated the development of structured workplace learning and have provided better linkages between informal and formal education and training. Frameworks have also allowed for the development of a common educational currency, which enhances the creation of education and training markets. This currency is buttressed by quality assurance processes, which in theory mean that no matter where an individual chooses to learn, they can be assured that the quality of the qualifications on offer are robust, widely recognized, and transferable. In this respect, Lauder and Mehralizadeh (2001) argue that New Labour is attempting to promote skill development through creating markets in education. Their argument is that New Labour has implemented a choice-based model of skill development. Accordingly, it is the decisions and judgements of consumers, operating in markets, that drive the provision of skill. However, they argue that a choice model is problematic, not least because it assumes individuals act rationally in their training decisions. In practice, the training decisions of many are made at some distance from the labor market, giving rise to inadequate decision making. Not only has this situation exacerbated skill mismatches, but it has also helped reinforce existing patterns of inequality. For example, the creation of markets in education has increased the incentive for providers to maximize enrolments, irrespective of demand in the labor market for the skills produced. The tertiary education sector has responded to the new environment by providing a myriad of course options, some of which appear to have little relevance in the labor market. This means that those wishing to enter fields that operate at a distance from the education and training system and who have limited or no understanding of the rules of advancement in operation are effectively duped into studying toward qualifications that are of little value. One effect of this is that rather than increasing open pathways into the labor market, frameworks instead increase the importance of having access to information about the labor market value of a given qualification on offer.

Recent developments, such as the Economic Transformation Agenda in New Zealand and the introduction of the Connexions Service in England, suggest that third way administrations in these nations no longer uncritically accept the choice model. For example, they do not accept that individuals are necessarily able to compete on equal terms, and they maintain that some groups need additional help before they can make rational decisions surrounding choice in tertiary education. Nevertheless, underpinning much of the third way's reform agenda in education is a belief that increasing consumer choice will lift standards of service as providers compete for clients. In turn, increasing the quality of education and training will help drive up innovative capacity and, ultimately, help build high-wage, high-skill labor markets. Thus, although important differences between New Zealand and England can be found, when analyzing the precise policies introduced and level of investment made, frameworks have been supported in both nations because they increase consumer choice and the formation of training markets. An important reason for this support is that economists favor the invisible hand of consumer choice as a way to facilitate social change. This is because it is perceived as far less messy and cumbersome than other methods (Hirsch 1976). The idea is that competitive market pressures reward those willing to adjust to new economic conditions and punish those who are not. For Hirsch (1976), an alternative means of facilitating change is through voice, or the notion that economic agents are influenced by persuasion, negotiation, and mutual education. Adjustment through voice is only likely to succeed if enduring bonds of mutual obligation and loyalty hold economic agents together. The institutions of the post–World War II period, such as comprehensive education, the welfare state, and trade unions, encouraged collectivism and, arguably, strengthened voice. The strengthening of voice is seen to have stifled innovation and to have undermined economic efficiency by, for example, limiting the ability of individuals to understand the moral requirements of the market (Marquand 1997). Put another way, an aim of free markets is to promote enterprise culture by increasing the ability of individuals to exit one market relationship and enter another. Indeed, maintaining the legitimacy of the economic system depends on reducing voice and strengthening exit. Distancing individuals from institutions that increase voice can help achieve this.

As implied, supporters of frameworks argue that they provide better, more flexible educational currency than those offered by earlier assessment systems. A key source of this flexibility is the way frameworks promote exit.

Frameworks promote exit by allowing individuals to move with relative ease within education and training systems, and from these systems into employment. They allow individuals to study toward qualifications at their own pace and in a variety of settings. For example, students can study on a part- or full-time basis at competing providers toward a single qualification. Frameworks also promote exit by opening up the provision of education and training to competing private and public sector providers. As central guardians of quality, the New Zealand Qualifications Authority, the Qualifications and Curriculum Authority, and related organizations in Ireland and Wales enhance the openness of the system by allowing competing providers to offer the same programs. In theory, elite institutions can no longer dominate the provision of higher education, because all accredited providers can offer education and training linked to frameworks.

The effect of frameworks on the creation of markets has been dramatic, with private and public sector provision expanding at a rapid rate in New Zealand and England. For example, providers of private tertiary-level training in both nations now offer a diverse range of qualifications and recruit their own students in direct competition with state sector providers. It is important to note, however, that concern over the quality of training offered by some private training establishments has led to reduced state funding in some areas in New Zealand. For example, the current government is closely assessing the relevance of courses offered by Maori tertiary education providers, such as Te Wananga o Aoteraoa. This review process has already led to the cancellation of several infamous courses, including one offering instruction in twilight golf and another in radio singing. However, in general, the private sector has increased its share of government funding and the full fee-paying student market. For example, before the establishment of the New Zealand Qualifications Framework and the creation of markets in tertiary education, preservice teacher education was provided by six state-funded teachers' colleges; at present, there are in excess of 30 public and private providers competing in the business of preservice teacher education.

Similar trends are apparent in England, where private training providers have also captured an increased share of the education and training market. As part of the competition for students, training providers have begun training people for a wider range of occupations than has traditionally been the case. For example, accredited training is now provided for new service sector occupations, such as travel and tourism, where a range of national certificates registered on the framework is available. Similarly, in the creative sector, a number of providers compete with one another for students interested in working in the film and theater industries.

Although frameworks have increased the involvement of private sector organizations in the provision of education and training, their effect on social outcomes has been disappointing, with high-status qualifications that provide access to selective tertiary programmes remaining the most highly desired. In England, the effect of the Curriculum 2000 reforms, which were designed to broaden the qualifications offered and increase the status of vocational learning by, for example, creating vocational A-levels, has been limited (Pring 2005; Waring et al. 2003). One consequence of this is that the participation of students from low-socioeconomic backgrounds in higher education remains low in both England and New Zealand.[9]

There are many other criticisms that can be leveled at frameworks. However, three should be sufficient to make the point that frameworks have yet to live up to expectations.

First, little evidence exists that employers, as a group, have embraced training as a key strategy for increasing their ability to innovate. The reasons for this are complex and varied but include the fact that low-skill routes to competitiveness remain viable (Keep and Mayhew 1999) and a desire on the part of employers to limit the need to pay workers more highly.

Second, although the existence of variations in the use of qualifications between fields means that caution is needed when interpreting the statistics, in general, there is little evidence that employers have adopted the job-competency approach to recruitment in all fields, or that framework qualifications can transmit the kind of information employers desire, particularly for work-bound school leavers (Wolf 2002). Nevertheless, some framework qualifications have enjoyed a strong following from employers. For example, the New Zealand Motor Industry Training Organisation continues to thrive on the back of a strong tradition of craft apprenticeships. However, other areas have struggled to get the support of employers. For example, attempts to set up an industry training organization (which is charged with developing industry-specific learning outcomes) for creative sector occupations have failed. Similarly, in England, a key justification for the development of the new Achievement Framework is that employers had not made the best use of older versions of the framework. Reasons for this included the complex nature of the existing framework, gaps in provision, and a lack of clarity in progression routes (Qualifications and Curriculum Authority 2004).

Third, it remains unclear whether or not increasing the ability to exit will have the effect of driving up the quality of provision, such that all learners receive the same quality of training. Naidoo and Jamieson (2005) argue that modularization of the curriculum, which is a key feature of frameworks, and the general marketization of tertiary education are likely to reduce the overall quality of provision by privileging certain kinds of knowledge. They argue that knowledge that is applied, transdisciplinary, and evaluated by external and internal stakeholders will be offered in the new environment. However, knowledge of this kind of knowledge does not necessarily equip students with the conceptual skills they need to become innovative. As the authors point out, not all institutions will be equally vulnerable to the effects of marketization and modularization, and those institutions at the top of the hierarchy will be best placed to resist. In this respect, although New Labour has successfully increased participation from lower-socioeconomic groups in higher education, their relative chances have altered little. Those from lower-socioeconomic groups "are as likely as not to be sorted into the lower reaches of the student population, obtaining lower end 'graduate jobs'. In other words, they are likely entrants to the two-year foundation degree courses" (Mayhew et al. 2004, 79). Although attempts to build frameworks and create open, or disembedded, markets in education and training continue, effort is also being exerted to build embedded markets in tertiary education. This issue is considered in the next section.

Networks, Skill Development, and Innovation

In addition to promoting skill development through building and expanding frameworks, governments in New Zealand and the United Kingdom are attempting to drive up innovative capacity by investing more heavily in research and development and in network formation. A criticism of the free-market reforms of previous neoliberal administrations is that they have not encouraged sufficient investment in research and development. For this reason, third way administrations are now attempting to create new research and development infrastructures that promote innovation. These new infrastructures are based on the formation of research and development networks and of innovation and employment networks. These networks can be seen in state support for research and development clusters specifically and in business-university links generally. It is acknowledged by third way governments that the state cannot itself form innovative networks. Rather, it can only establish the conditions under which networks can take hold. As a consequence, the state sees itself as a strategic broker of networks (Department of Trade and Industry 2007a).

It is not possible to fully document the range and scope of the measures in the United Kingdom. However, to give readers a sense of the new measures, collaboration between knowledge generators and users of knowledge has been mandated by the Department of Trade and Industry, and the research councils are now required to establish measures of collaboration so that progress can be assessed (Department of Trade and Industry 2003). As argued in more detail in Chapter 7, similar developments can be seen in the formation of knowledge transfer networks, which are also overseen by the Department of Trade and Industry. Knowledge transfer networks are overarching networks operating within specific fields. The networks bring together a variety of actors, such as knowledge generators, finance companies, businesses, and other brokers to share knowledge and enhance the commercialization of innovative knowledge. Knowledge transfer networks are also designed to increase the pace of knowledge transfer and help build high-performance workplaces (Department of Trade and Industry 2003). Although the very notion of what makes up a high-performance workplace continues to be a source of debate, most conceptions of the term combine ideas about the use of advanced technologies, modern processes, and management to increase returns from the production process. As the argument goes, high-performance workplaces are also rich in forms of social capital such as team working, employee involvement in decision making, communication, and trust.

As is the case in other areas of education, the New Zealand Labour government has been much less active than New Labour in England in supporting network formation. Nevertheless, network formation is now high on the Labour-led government's list of priorities. As part of forging a third way, the Labour-led Coalition believes that it should work in partnership with providers of education and training. The effect of this new approach to policy can be seen in a number of settings. For example, in preservice teacher education, recent measures are designed to help form professional networks. Similarly, in the schooling sector, the Extending High Standards in Schools initiative is an attempt to improve the quality of teaching by providing effective schools with additional funding so they can share their teaching and learning strategies with the broader educational community (Strathdee 2007). As noted, the general idea is that governments can work as strategic brokers, bringing groups together and investing where the risks to business of doing so without the state's assistance are too great.

Network formation is also an integral part of Labour's Economic Transformation Agenda, which aims to boost New Zealand's capacity to innovate by boosting skill development and encouraging greater collaboration and cooperation in research and development. As is the case in the United Kingdom, to facilitate this, the Labour-led government has established itself as "leader, partner, facilitator, and broker working with other sectors to get results" (Maharey 2002, 1). The recently introduced Centres of Research Excellence, which have been established in areas of major importance to New Zealand's economic and social development, are a good example of this. The Centres of Research Excellence operate collaboratively across tertiary institutions, and many have links to Crown Research Institutes and sources of private sector funding. Similarly, industry-led research consortia were introduced in 2002 with the aim of establishing collaborative research. Consortia always involve at least two users of research, for example, businesses and at least one research provider, such as a Crown Research Institute or a university. They may also include overseas entities. The research consortium participants must provide at least 50 percent of the cash requirements of the projected research programs. Government funding is also limited to a maximum of seven years. Although the associated discourse is much weaker and the level of investment much lower in New Zealand than in the United Kingdom, network formation is also seen as being important in the creation of high-performance workplaces (Report of the Workplace Productivity Working Group 2004). Finally, network formation is a key aspect of the new model of funding tertiary education. The idea is that providers of tertiary education will work closely with each other and with the government to develop a network of provision (see Chapter 8).

For a partial explanation of the growing government interest in network formation, we can turn to researchers in economics and related disciplines, who have long argued that globalization has exacerbated the divide between core and peripheral regions and between competitive and less competitive regions (Howells 2005). A central aspect of this divide is the emergence of regions of innovative milieux or clusters, in which geographical proximity, informal relationships, and networks between firms and knowledge generators, along with the presence of shared values and understandings, combine to drive up innovative capacity. These networks and relationships are important aspects of what Amin and Thrift (1994) refer to as institutional thickness. It is the institutional thickness, conceptualized as the depth of social capital operating in an environment of investment in knowledge creation, that drives up competitive advantage.

Innovation and Social Capital

When assessing the connections among knowledge, innovation, and networks, it is useful to briefly review how innovation arises in knowledge economies. In the 1990s, scholars such as Freeman (1997) built on the insights of Schumpeter (1934) to theorize the relationships among innovation, technological change, growth, and trade. In turn, Schumpeter's arguments can be traced back as far as Marshall's nineteenth-century notion of industrial atmosphere, which was seen to improve innovation through a variety of mechanisms, including reducing transaction costs and facilitating knowledge transfer (Marshall 1961). Marshall's and Freeman's insights have influenced our conceptualization of the knowledge economy in key ways. First, they have highlighted the role played by technological change as a key driver of economic growth in the current era. Of significance here is the advent of microelectronic technology and related developments in such areas as computer technology, biotechnology, and information technology. The emphasis on technological change as a driver of innovation in high-skill, high-wage economies underscores the emphasis on techno-scientific knowledge in the knowledge economy. Second, the relationship between techno-scientific knowledge and innovation leads to a conceptualization of the knowledge economy as a national system of innovation. Accordingly, it is a nation's public and private sector institutions—and their activities and interactions—that both create and diffuse new technologies (Freeman 1997). Such interactions, seen as untraded goods, go beyond individual firms and are based, for example, on forms of social capital for which no market mechanisms exist and that are consequently difficult to price.

Freeman's (1997) insights have been taken up by global organizations. As the Organisation for Economic Co-Operation and Development explains, "the configuration of national innovation systems, which consists of the flows and relationships among industry, government and academia in the development of science and technology, is an important economic determinant" (Organisation for Economic Cooperation and Development 1996, 4). A proper analysis would extend the idea of systems of innovation to include those relationships and networks operating at international level. Of course, the source of the innovations will exert an important effect on the nature of the networks in operation and will vary across fields. Although they do not offer such an account, the Department of Trade and Industry and the Tertiary Education Commission have adopted the idea that innovation is encouraged through the creation of knowledge networks. This move is underpinned by the notion that new knowledge is social in character. In other words, new knowledge is created by individuals working together, rather than by individuals working in isolation. For example, the term learning economy describes innovation as being fundamentally social in character and resulting from collaborations between individuals rather than from individuals working alone (Doloreux 2002).

In its many variations, the notion that innovation is fundamentally social builds on the ideas that all knowledge cannot be concentrated in one individual mind and that no single mind can know in advance what knowledge will be of value in the future. For such reasons, the creation of new knowledge is fundamentally a social process involving social capital in the forms of cooperation, trust, and the presence of shared language codes, values, and norms (Archibugi and Lundvall 2001; Nahapiet and Ghoshal 1998). Moreover, according to some, the most innovative firms have access to a strong local knowledge base and international sources of knowledge. Access to knowledge and other capital such as finance and entrepreneurial capacity are key resources that allow innovative firms to maintain their competitive advantage in the global economy (Simmie 2003).[10]

As argued in previous chapters, a further reason why innovation is created through social mechanisms is that much innovative knowledge is tacit in character. Such in-house knowledge is often difficult to transfer because it is not easily codified. This means it resists codification in manuals and other texts and, critically, cannot be easily embodied in qualifications, standards of competency, or frameworks (Strathdee 2005b). Its transfer depends largely on labor mobility and informal, face-to-face information sharing via social networks. Indeed, once innovations have become sufficiently standardized and codified such that they can be embedded in qualifications, they are arguably no longer a source of advantage. This is because competing firms then have access to them through recruitment and training strategies. In this respect, open networks do not provide advantages to any groups or individuals, as all who want to can access the network and derive precisely the same sort of information or benefit. Once innovative knowledge has been transmitted through labor mobility to competing firms, its value as a source of competitive advantage decreases. Thus, access to tacit or noncredential knowledge is essential if firms and nations are to create high-wage forms of employment.

Although it is important to stress that it is not the only requirement, from this discussion it should be clear that innovation, because of its social character, is likely to come from the creation of new networks and the creation of linkages between research and development organizations, particularly universities, which are directly linked to innovative firms. These linkages provide a critical method of knowledge transfer. That universities have a central role in driving innovative capacity was also a key theme of the recent Lambert Report into university and business links (H.M. Treasury 2003). By arguing that knowledge creation and transfer require human interaction, the Lambert Report amplifies themes found in the social capital literature.

Networks and Social Exclusion

Some see the development of networks as making social sense. Here, the argument is that network creation holds out the possibility that universities and the communities in which they are situated will form new links. In turn, this is thought to increase the likelihood that civic traditions will be enhanced and democracy thickened (Peters and May 2004). However, the analysis presented above challenges this idea and argues that those distanced from such forms of social capital and the means to decipher emerging language codes (Szreter 1998) are likely to be excluded from knowledge flows and other networks, and hence be excluded from participating in embedded markets. Thus, network creation establishes new forms of social exclusion based on access to positional knowledge and possession of social and cultural skills needed to create and maintain networks. In sum, markets are created, but their successful functioning depends on, and indeed privileges, embedded relationships.

When considering these comments, it is important to remember that the rules of advancement vary from field to field. In some fields, innovation is not derived from knowledge but from other forms of human capital, such as aesthetic skills. Moreover, exploring the relationships among training, networks, and employment in the new economy highlights the complex relationship between embedded and disembedded markets. In this context, it is worthwhile noting Blair's (2001) research, which shows that, similar to other fields, securing employment in the creative sector depends on gaining and maintaining insider status; the competition for a place in the industry is far from open, with those who possess social capital best placed to succeed.[11] This research suggests that initial employment in the industry is gained through kin- and community-based networks. Subsequently, professional networks and participation in teams is of greater importance (Blair 2001).

Blair's (2001) research did not specifically explore the relationship between qualifications and recruitment; nor was the relationship explored over time. However, in fields where skills needed cannot be obtained informally, or where individuals lack the means to gain them informally, it is likely that some kind of formal training will be required. In such instances, it is likely that both social capital and human capital combine to help create pathways into the labor market. Finally, it is also likely that, as fields develop, the mix of resources needed for advancement change. For example, the need for new recruits to have formal educational credentials in emerging fields, where skills from a number of different areas of the labor market may be being combined in new ways, may differ from the need in more developed, or older, fields that have established training arms.

Conclusion

To date, there has been little research into how skills policies and strategies of third way administrations are creating different kinds of markets in education and training. This chapter has attempted to address this weakness and has shown that third way administrations appear committed to creating both disembedded and embedded markets. In terms of disembedded markets, the chapter has examined the fact that the creation of frameworks can be seen as a means of facilitating innovation through increasing skill development and diffusing knowledge throughout the economy. Frameworks are also touted as a method of breaking down forms of social closure that have, until now, limited individual progression through the labor market. In terms of embedded markets, current efforts to promote innovation and knowledge transfer through network creation have been explored. Network creation is needed to support the exchange of knowledge and ideas between individuals and institutions. Such untraded goods are vital to innovation of a kind that creates high-skill/high-wage forms of employment, but they are relatively poorly understood. For example, we know little about how networks contribute to competitive advantage in different fields.

Although the creation of frameworks is presented as complementing network creation, the analysis presented here suggests that these strategies have different effects. The creation of frameworks has increased participation and achievement. However, their contribution to innovation and social inclusion is likely to be more limited. Participants in different kinds of markets in education and training are likely to have different labor-market outcomes, with those in embedded markets enjoying greater access to employment in the knowledge economy. The creation of networks heralds the introduction of new forms of social exclusion based on the principle that innovative knowledge is likely to be privatized in the new entrepreneurial environment. Those attending institutions poorly embedded in the social infrastructure will obtain qualifications, but their value in the labor market will be lower than those obtained from institutions embedded in the social infrastructure. In sum, the chapter's premise is that contrasting pathways leading to different destinations in the labor market are being created by third way administrations. However, one glaring limitation of the analysis is that it does not consider how the rules of advancement might differ within or between fields. The next chapter, which focuses on the creative sector, pursues these issues in greater depth.

CHAPTER 5

Innovation and Networks in the Knowledge Economy
The Case of the Screen Production Field in New Zealand

Introduction

Each day during the last weeks of July 2006, Vinnie Duyck erected and inhabited a small tent outside Peter Jackson's Weta workshop in Wellington, New Zealand. Vinnie aimed to demonstrate the strength of his desire to be employed at Weta, the place where Peter Jackson and his team produced the *Lord of the Rings* film trilogy. A recent graduate of Oregon's Pacific University (majoring in the Arts), Vinnie had travelled to New Zealand to immerse himself in the movie-making culture and to find work in the screen production field. However, he had been told that there was no work available at Weta at the time. While he waited for an opportunity to work in film, Vinnie was working as a barista in a café across the road from Weta. Armed with a tertiary-level qualification, a "large dose of determination and an example of his model-making prowess," Vinnie hoped to signal to Peter Jackson that he had the raw talent needed to work in Weta. As he put it, "I'm going to stand outside here, till I stand out inside" (Mulrooney 2006, 7).

This chapter explores the emerging screen production field in Wellington in relation to network formation. In New Zealand, screen production is primarily based in Auckland, which dominates the production of shorter works for television, and Wellington, which concentrates on the production of longer works, particularly film. Specifically, the chapter assesses how managers solve their human resource challenges and uses this information to shed light on processes of social reproductionadvancement that occur in this field. As argued in previous chapters, processes of advancement are different in different fields, and a better understanding of these will aid researchers and policy makers. In exploring the behavior of managers, the chapter also considers the role of tertiary education in mediating the competition for advancement in the screen production cluster. As noted, the creative field is interesting, both because it is held up as evidence of New Zealand's ability to compete in the global economy in ways that create growth in knowledge work and because it is an area in which the state has claimed that its interventions have made an important, if not critical, contribution to growth in knowledge work.

The chapter proceeds through the following sections: section one briefly backgrounds the case study by describing work in the screen production field in New Zealand. Section two describes the labor market in creativity and assesses the role of enterprise culture as part of a wider process of establishing and policing the rules of advancement in operation in the screen production field. To provide some focus to the analysis, after describing the method of data collection, the following sections draw on interview data with managers within the screen production field. These data are used to assess how the labor market for creativity functions and to assess the relationship between investment in education and training and innovation.

Work in the Screen Production Field in New Zealand

The possibility of working in the creative field has captured the imagination of an increasing number of people similar to Vinnie. Many have flocked to courses linked to the screen production field. For example, although dated, survey evidence from the United Kingdom suggests that increasing proportions of people aspire to work in the creative field as artists or designers. In 1994, one out of 64 applicants for university places wanted to study either design or fine arts. In 1999, the figure was one in 19 (Heartfield, cited in Tepper 2002).

Unfortunately, no information on the number of students undertaking training linked to employment in the creative field is available for New Zealand. However, if growth in the number of providers is any basis to make a judgment, tertiary-level training linked to the creative field has become increasingly popular in recent years. Before the development of the National Qualifications Framework and the creation of markets in education (see Chapter 4), there were very few providers of education and training working directly to provide skilled workers for the screen production field. Former government departments contributed in a minor way to accredited skill development (particularly the New Zealand Broadcasting Commission).

They did this by providing traineeships in areas such as film and television production.

Since the days of the New Zealand Broadcasting Commission, the provision of training in film and related areas has expanded dramatically. There now exist in excess of 25 providers of training listed on the New Zealand Screen Council's[1] Web site that provide training for those wanting to enter the screen production field. The list comprises a diverse range of providers and includes private training providers, polytechnics, and universities. These providers offer a wide range of courses in all aspects of film, including sound production, film production, acting, costume design, and the like. For example, Victoria University of Wellington offers undergraduate and postgraduate degrees in both film and theatre. Some qualifications offered by these providers are registered on the National Qualifications Framework, such as the National Certificate in Film and Television, and comprise various unit standards (or learning outcomes). Others offer unit standards linked to their own diploma- or certificate-level qualifications or, in the case of the universities, their own degree qualifications accredited by the New Zealand Vice-Chancellors' Committee.

The Labor Market in Creativity

The potential of the creative field as a source of employment and economic growth has not escaped the attention of policy makers and politicians in New Zealand and the United Kingdom. As commentators have noted, growth in the creative field has provided a way to promote New Zealand as a global center of creative capitalism where the raw (or native) creativity of individuals is combined with economic capital, advances in technology, investment in education and training, and the country's natural beauty to produce growth in knowledge work (Jones 2005). When combined with the role of the state in supporting the creative field through such ventures as the establishment of the film production fund (which provides limited funding to produce feature films) and other measures such as offering tax breaks to large productions and increasing the supply of suitably skilled labor through training reform, a picture of the state influencing the nature of—and growing—the screen production field emerges. Indeed, state support has been a critical ingredient in establishing a screen production field in New Zealand (Jones 2005) and it has been identified by the government as a key field to foster growth in knowledge work across the national economy. In the year that ended February 2004, the screen production industry (comprising about 2,000 enterprises) provided employment for nearly 6,000 people and generated total sales in excess of NZ$2 billion (New Zealand Screen Council 2005).

As suggested earlier, it is difficult to provide a clear and concise definition of the creative field and to determine where the boundaries between it and other fields lie. The creative field is generally thought to include industries that design, produce, and deliver creative content, as well as industries that integrate technical skills with creative and artistic talent. In New Zealand, the creative field tends to be most closely associated with the screen production field, but other areas, such as fashion and design, are gaining prominence. One reason that the screen production field dominates the field in New Zealand is that it appears to be experiencing the greatest growth. In 1995, Peter Jackson commented, "we don't have a feature film industry here" (Campbell 1995, 20). Although such comments may be accurate when New Zealand's screen production field is compared with that of other countries, it probably under-states the case. Until 1978, feature film production was infrequent, with no continuity of production and little, if any, investment finance. After this date, intervention by the state, which established the New Zealand Film Commission, led to growth in feature film production for the free-to-air television market. This also testifies to the importance of the state in helping to form the field. Since the 1960s, the state had controlled the production of both television programming and, to a lesser extent, film through its monopoly on transmission. The production of programs and films tended to be under-taken in house, and there was little independent film production. With the establishment of the New Zealand Film Commission, this began to change. Independent producers were able to gain limited financing to support their productions for the free-to-air television market. These producers were also able to benefit indirectly from state support by recruiting new workers who had been trained by the New Zealand Broadcasting Commission. Following the establishment of New Zealand on Air in the 1980s, larger independent film productions were commissioned. Several of these gained success in the Cannes Film Festival, helping pave the way for the business in New Zealand to lose its status as a cottage industry (New Zealand Screen Council 2005).

In recent years, investment in film production has increased dramatically. In 1994, the value of total production was estimated to be approximately NZ$150 million. By 2001, production of films such as *The Lord of the Rings*, among others, boosted spending to an estimated NZ$527 million before dropping back to a little over NZ$200 million in 2004. A number of new films commencing production at this time, such as *King Kong*, gave the New Zealand Film Council reason to estimate that investment in the screen pro-duction would reach almost NZ$600 million in 2005 (New Zealand Screen Council 2005).

Increases in the level of the financial investment in screen production have been mirrored by increases in employment in the field. However, because much of the work in the field is organized around short-term contracts, accurately measuring the total level of employment has proved problematic. The best figures estimate that about 5,900 people were employed in approximately 1,800 enterprises in the field in the year ending 2004 (New Zealand Screen Council 2005). To speculate, it is likely that employment is polarized between relatively few large enterprises, which employ many people, and many smaller enterprises (or contractors) comprising one or two employees.

Data on the number of individuals employed directly in the field and data on direct investment in the field only tell part of the story. For example, they does not account for those employed in downstream activities, which also generate significant economic activity. These include boosting investment in media technology, tourism promotion, enticing New Zealand talent to return home, and building New Zealand's global profile. In addition, as noted above, growth in training for work in the screen production field has also boosted employment. As a result of this and other activity, the Department of Statistics has estimated that the screen production industry generated over NZ$2.6 billion in revenue (Department of Statistics 2005). Such data help understand how rapidly the field has grown. Nevertheless, even with such evidence, commentators point out the "hoard of the Rings" will be difficult to measure (Calder 2003).

Part of the value to the state of advancing New Zealand as a creative economy is that it provides a way to give emphasis to New Zealand's ability to be competitive in the global economy in ways that increase employment in knowledge work. In the absence of convincing evidence that its policies in education and training—and in other areas of the economy in general—have delivered the promised growth in knowledge work, the screen production field has arguably provided the state with a much-needed source of legitimacy for the interventions it has made. Indeed, creative successes, such as the films *The Lord of the Rings* trilogy and the *Chronicles of Narnia: The Lion, the Witch and the Wardrobe*, are held up as evidence that with the support of the state, New Zealand can compete with the best in the world and create knowledge work. As New Zealand Prime Minister Helen Clark put it in the case of *The Lord of the Rings*: "Set against the spectacular and diverse New Zealand landscape, *The Lord of The Rings* trilogy has the potential to be a major tourist promotion and investment tool for years to come, by highlighting the country's natural beauty and the creative talents of its people across a wide range of knowledge-based industries" (Clark n.d., 1).

Jones (2005) argues in more detail that the involvement of government in the local screen production field has grown dramatically in recent years, from relatively small, but not insignificant, amounts of funding via New Zealand On Air for free-to-air television production and related initiatives, through to large funding packages designed to promote spin-offs from major productions, such as the *Lord of the Rings*. In related developments, the Labour government even appointed at one point a Minister of the Rings to maximize the benefits that could be derived from producing *The Lord of the Rings* trilogy ("Minister of the Rings" 2001).

In addition to providing much needed legitimacy for its interventions, increased investment in the screen production field has formed a key plank in Labour's nation-building project. Accordingly, it is part of the current government's cultural recovery platform in which government investment is linked to the formation of a national brand. Indeed, through the melding of New Zealand's cultural, creative, and natural assets with their commercial interests in the increasingly global marketplace, the state is attempting to modernize New Zealand's national identity to accommodate new economic conditions. In these new times, notions of enterprise and creativity are being combined to form a new field. In this field, elites appear to be those who can form new, hybridized identities that allow them to become boundary spanners and promote growth in knowledge work. By bringing together their creative talent, New Zealand's natural assets, the latest technology, and their international contacts and expertise, Jones (2005) argues that elites are able to develop new cultural products in ways that appeal to consumers globally. As suggested, the evolution of the field is driving changes in demand for creative skills. The next section looks at this issue in detail.

Research on the creative field suggests that much of the work in the field is project based, with workers moving from employer to employer with relative frequency. Moreover, much of the work in the field involves participating in teams, which are formed for specific projects (de Bruin and Dupuis 2004; Dex et al. 2000). In many respects, then, developments in the screen production field provide evidence in support of Castells' (1996) prediction of what the labor market generally will look like in the future. Castells argues that networks provide access to sources of various capitals and access to technical know-how, management, and information. In Castells' view, networks operating globally and locally will increasingly characterize production and be linked to competitiveness. Indeed, in his view, network enterprises are emerging as the new standard for business. Thus, enterprising firms are networked with other firms. In turn, the prosperity of individual workers is also increasingly tied to participation in networks.

Whether or not Castells' vision of the network society will become a reality as the economy develops will remain a question for future empirical research and testing. However, relatively little is known about the rules of the competition for the advancement in the field of screen production, or in the way various forms of capital are operationalized in the field. One source of questioning arises from the fact that creativity is very difficult to teach and to assess through bureaucratic methods, such as on the basis of educational qualifications and open interviews. This suggests educational qualifications may play a limited role in the competition for advancement in this field. Difficulties assessing and teaching creativity imply that other ways of signaling competency are required for entry into, and advancement throughout, the field.

As argued in this book, economic and social changes are forcing a refiguring of the rules of advancement. In this respect, in the past, creative workers (or artists) were seen as being separate from business people and from entrepreneurs. They were more or less seen as distinct entities, having different histories and contrasting ambitions. In the contemporary period, developments within the creative field and intervention by the state are contributing to the amalgamation of artists and entrepreneurs, such that providers of art need also to be entrepreneurial, or at least make links with sources of finance (Wu 1998). In this respect, a key development in the field is the way creative workers (and providers of creative works, such as art galleries) are forming alliances with sources of finance to achieve shared goals. This is also part of a reformation of the field as the boundaries between art and entrepreneur are being redrawn.

Linking growth in the creative field to broader social and economic objectives has allowed the state to legitimate its intervention in the field and provide a way to highlight New Zealand's distinctive contribution to the new global economy. Consistent with third way politics, investment by the state is a central aspect of a cultural recovery package (Tizard 2001). Providing more funding to the arts, culture, and heritage fields will stimulate economic growth and increase demand for knowledge workers. As Ruth Harley, Chief Executive Officer of the New Zealand Film Commission, argues, the new hybrid creative industries provide an opportunity to reconcile the cultural and economic benefits of investing in the creative field and can be employed to rework our national identity.

> Cultural industries such as film and television, fashion, multi-media, music and tourism are transforming New Zealand's economy. Our commercial interests are indissolubly linked with our cultural interests. There is no place in the new economy for the type of thinking which sees a disjunction between the business world and the art world. Cultural industries are based on national identity. National identity is key to creating a unique positioning for our goods and services. Take film for example. It creates culture, builds identity and markets that identity to the world. Film is important not just as a potent advertising medium for New Zealand; not just as a way of creating and personifying our country as a brand in all its diversity; not just as a high growth, high

margin knowledge based business. It is all of these, but it is also as a statement to ourselves. It is a central ingredient in constructing our identity for ourselves, as a lever to help New Zealanders get the confidence and boldness to foot it aggressively on the international stages. (Hartley, cited in Jones 2005, 7)

In the contemporary period, then, the merging of art and entrepreneur has led researchers to describe the creative industries as "occupying a relatively new interdisciplinary space" (Jeffcutt and Pratt 2002, 225). In this context, innovation arises through the merging of creative/cultural/artistic talents with economic/business/industry interests and acumen. As Jones (2005) argues, by rebranding the cultural industries as distinct from economic interests—as the creative industries—the state has played a critical role in this process. Competitive advantage in the global economy can be gained through the production of novel or unique cultural products, which cannot be produced anywhere else in the world (Jeffcutt and Pratt 2002). In this context, the New Zealand government has been quick to link successes such as *The Lord of the Rings* to the nation's natural resources, which are claimed to be unrivalled anywhere in the world.

The establishment of enterprise culture by the state, as a guiding principle, is central to the formation of the creative field. It is also an important aspect of the overall attempt to instigate new rules of advancement operating within the field of screen production. According to Dean (1999), for example, "governmentality" refers to an "assemblage of practices, techniques and rationalities for the shaping of the behaviour of others and of oneself" (Dean 1999, 198). From this perspective, entrepreneurial culture can be seen as a technology of government designed to engage individuals in a process of identity re-formation by encouraging them to act on those "attitudes, affects, conduct and dispositions that present a barrier to . . . them participating in the labour market" (Dean 1995, 572). As such, the discourse associated with creative industries extends the neoliberal enterprise culture of the 1980s and 1990s (du Gay 1991) to include a duty on the part of the individual to be entrepreneurial through being creative. To draw relevance from McRobbie (2002) to the New Zealand case, through developing New Zealand's own native or raw creative capacities, it is believed that individuals can create their own jobs in the knowledge economy and, in turn, employ others. Thus, new forms of identity are opened up, such as the creative entrepreneur. Indeed, the idea that work in capitalism itself can be creative is itself opened up. This represents an amalgamation of psychology, which has long emphasized creativity as a general cognitive attribute in humans and business managers, who have linked creativity to innovation and, hence, competitive advantage. Using this model, creativity is not something that emerges spontaneously;

rather, workers can be organized to increase creativity. Now there exists a welter of business techniques that harness the insights of psychology to increase creativity in the workplace (Rickards and Moger 2000). In this sense, creativity is not only a property of individuals but also a property of systems, and especially networks of individuals.

When considering the relationship between education and the competition for advancement through education in the screen production field, it is necessary to note the way the promotion of enterprise culture has emerged as part of neoliberal reform programs in England and New Zealand. It is necessary to describe the rise of enterprise culture briefly here because it has been linked both to growth in the creative economy and to the introduction of new rules of competition for advancement.

A proper understanding of the rules of advancement operating within fields can only be gained through research and testing. However, the emphasis accorded such traits as raw talent and entrepreneurial spirit in the discourse associated with the creative field alerts us to the possibility of a minimal role, in practice, for education and training in the competition for advancement in the creative field (although the emphasis on the importance of education and training in the wider discourse remains). As outlined above, proponents of the technocratic-meritocratic perspective maintain that the expansion of education and the increasing use of bureaucratic recruitment methods (both of which are driven by development of new technologies and the competitive pressures of capitalism) would increase both efficiency and fairness in recruitment. In this respect, a key aim of qualifications frameworks is to aid in the production of institutional-based forms of trust, which enhance the construction of open competitions for advancement and bring the provision of education and training into closer alignment with the needs of the labor market.

Despite the emphasis in discourse associated with the creative field on raw creative talent and entrepreneurial skills (as discussed earlier), training in "creativity" has grown steadily in recent years. However, as noted, there remains considerable doubt about the relationship between this training and employment in the screen production field. One source of doubt rests in the way creative capacity is signaled to others. The official view is that educational qualifications signal the skills and qualities needed to be effective in the labor market and that employers use this information as a basis for selecting new workers. However, the ability of qualifications to signal creative capacity remains an open question.

Earlier research conducted in New Zealand (de Bruin and Dupuis 2004), as well as research from the United Kingdom (Dex et al. 2000), has helped us to understand more about how employment is facilitated in the field.

This research highlights the importance of networks as a means of gaining employment in the screen production industry. For example, the workers interviewed by de Bruin and Dupuis (2004) reported that the project-based nature of employment in the field meant that employers often had a limited time to employ team members. Although de Bruin and Dupuis do not express it as such, contacts within the recruiters' networks were often used to find workers because they provided a reliable source of information about workers (see Blair 2001 for research on the British case). However, on the basis of their data, de Bruin and Dupuis reported that nepotism meant that recruits were not always suitably competent.

The relationship between training in creativity and employment in the screen production field can be questioned further on the grounds that the level of skills needed to be effective in the field varies dramatically. Indeed, the field is diverse, and the little research that has been undertaken on the issue in New Zealand suggests that much work in the field is poorly paid and requires little, if any, training (de Bruin and Dupuis 2004).

Last, despite the state's belief that employers will find qualifications produced by the Qualifications Framework in New Zealand of increased value to all users, evidence suggests otherwise. For example, despite its rapid growth, there is no industry training organization representing screen production. In fact, a former industry training organization, Film and Electronic Media, collapsed in 2000. This suggests that employers in the field either see no need for a systematic approach to training or had no faith in the ability of the Film and Electronic Media Industry training organization to deliver it. The approximately 350 unit standards established by this industry training organization were deregistered by the New Zealand Qualifications Authority on May 12, 2006. In contrast to some overseas countries, where new workers can gain skills via an apprenticeship, there are no similar training schemes applicable to the screen production field in New Zealand. Instead, those interested in gaining formal training must attend registered training providers, including private training establishments, polytechnics, and universities.

Concern about these and other aspects of work in the creative field has led critics to argue that the state's promotion of enterprise culture is part of a broader strategy designed to encourage people to adjust to the realities of working in an area in which much work is inherently insecure and relatively poorly paid. As part of creating social conditions supportive of the emerging field of screen production, the discourse associated with the creative field has emphasized that, to remain in work, individuals need to manage themselves in new ways, particularly by being entrepreneurial (du Gay 1997).

It is also important to remember that growth in project-based work and the related development of workers' identities has also been encouraged by policy changes introduced by successive governments. Of significance here is the way funding provided by the state is specifically intended to support one-off productions, which have definite completion dates. Although it would be a mistake to see the independent field as made up entirely by freelance employees, much of the work directly in the field and in downstream industries relies on the successful financing of one-off projects, such as *The Lord of the Rings* trilogy. Thus, although the major television channels continue to offer secure employment for some, much of the growth in the field has arisen from increases in the number of independent producers who tend to employ freelance workers. It is well known that this characteristic of the field has contributed to uncertainty and a heightened sense of personal risk for workers in the field. Workers need to adopt a range of strategies to manage their flexibility. These include diversifying their income portfolio (including relying on their partner's income during lean periods), collecting information, and building networks (Dex et al. 2000). Although figures for New Zealand are not available, estimates from Britain, which has experienced similar political and economic changes, suggest that between 1979 and 1990, the British Broadcasting Commission shed 12,000 permanent jobs, followed by a further 19 percent of jobs between 1990 and 1993 (Antcliff 2005). The loss of these jobs was partially compensated for by growth in the number of independent workers, and it has been estimated that over 50 percent of workers in the television industry were freelance by the early 1990s (Woolf and Holly 1994).

To date, research has stressed the importance of individuals managing their lives to cope with the required flexibility, and it has highlighted the importance of networks in the field. However, a full understanding of both the relationship between tertiary education and employment in the field and the way managers build and use networks to satisfy their need for labor power is lacking. The remainder of this chapter is devoted to addressing this question through exploring the labor market in the screen production cluster that has developed in Wellington.

Method

Senior managers working for larger screen production houses, owner-operators of smaller independent production houses, and providers of training linked to the field were all interviewed as part of this study. The majority of informants were based in Wellington. However, a small number were based in Auckland. Although these informants resided outside of the Wellington region, they worked in areas not practised by the interviewees in the Wellington cluster. For example, they completed specialist postproduction activities.

One aim of the interviews was to map the human resource practices across the field, with a view to assessing the relationships between tertiary education and the screen production field. The New Zealand Screen Council's Web site provided details on employers in the Wellington cluster. These data was used to identify potential informants. From here, a grapevine recruitment strategy was used to recruit additional informants. In total, 15 managers were interviewed (13 from Wellington and two from Auckland). This represented the majority of enterprises in the Wellington cluster as well as some smaller organizations that serviced it (but were located outside of it).

A qualitative methodology was employed based on semistructured interviews. The interviews were designed to gather a range of data including that pertaining to the role played by managers in recruiting skilled workers and their views on the training of screen production workers. As part of the ethical considerations that underpinned this study, all participants were told that the purpose of the research was to explore the practices and strategies they used to recruit new staff and about the relationship between their enterprises, skill, and the tertiary education field. Participants were guaranteed anonymity, informed that they could withdraw from the study at any time without reason, and told they could withdraw any data already provided. Pseudonyms have been used to protect the identities of all participants and their organizations. The interviews were audio recorded and lasted between 30 and 60 minutes. As soon as practicable after interviewing, relevant data (i.e., data that were considered central to the enterprises and individuals, satisfying their human resource requirements and including those related to skill demand) were transcribed. Data analysis involved identifying the key themes and experiences of each interviewee. To improve the readability of the data, some text has been removed and some added. In this respect, an ellipsis (. . .) indicates text that has been removed, and square brackets ([text]) are used to indicate text that has been added. An early draft of the case study was returned to the participants so that they could attest to its accuracy. Respondents suggested a number of minor changes be made to the background information and to the reporting of the data. For example, one manager wanted the data altered from "manager" to "production manager" to better reflect the kind of work a recruit was undertaking. All changes that were suggested by the informants were made. It is important to acknowledge that this study explores the behaviors of a small number of participants who work in particular locations. These factors should be borne in mind when the study's findings are generalized to other settings.

The Creative Cluster in Wellington

As argued throughout this book, the official view is that investment in tertiary education and training is critical to building national systems of innovation. More recently, this view has been supplemented by the view that investment in knowledge creation is not in itself sufficient to promote innovation—methods of knowledge transfer are also required. Because knowledge transfer is critical to innovation, governments in New Zealand and elsewhere have designed policies to bring creators of innovative knowledge and enterprises into close association. Previous research conducted in New Zealand and the United Kingdom has highlighted the importance of social networks in recruitment processes operating in the creative field (de Bruin and Dupuis 2004; Dex et al. 2000). Although this research has improved our understanding, as with the official discourse (and, for that matter, some of the academic literature), there is a tendency to portray the creative field as homogenous, with similar rules of advancement in operation throughout the field. For example, Jeffcutt and Pratt (2002) suggest that the merging of art and entrepreneurship is producing new forms of work and new worker identities. Although at a general level it is possible to agree with the thrust of Jeffcutt and Pratt's analysis, it fails to take into account differences that might exist across time and space. In a similar way, the rules of advancement in operation within fields are neither static nor unchanging. Rather, they undergo ongoing processes of formation and reformation as fields develop and evolve in response to ongoing social and economic changes and in response to intervention from the state. As a result of these changes, the value of different forms of capital also changes.

The key themes that emerged from the data were the diverse nature of the field, the selection and recruitment of new workers, sources of innovative knowledge, and links with tertiary providers. These themes are discussed in turn.

Diversity of Field

Early on in the data collection, it became clear that the screen production field is diverse, with different kinds of productions geared to meeting the needs of a variety of audiences. One consequence of the diversity in the field was that the human resource requirements varied. For example, when compared with production for television, there were relatively few permanent positions in film, with most of the employment on offer being only available in the form of short-term contracts. At the same time, across both areas, fluctuations in production also meant that managers needed to alter their human resource practices. For instance, with large films, such as *The Lord of the Rings*, most sound engineers were employed on the one film. In such instances, local capacity was stretched, and managers needed to dig deeper into their social networks to find staff.

Although networks in various forms played an important role in facilitating labor market relationships in the field, the data also suggest that the way they do this differs both within individual companies (e.g., as they cope with fluctuations in production) and between companies that meet the demand for specialist services. The diversity of production meant that some kinds of employment were relatively secure, whereas other forms relied more heavily on short-term contracts. In the case of television production, for example, team leaders tended to come from within the core of the respective companies and also bore responsibility for identifying and contracting team members. In contrast, the human resource managers reported that the kinds of skill sets needed were different in film, where team skills were allocated higher importance than the skills required for work in other areas, such as television. Because of such variations, different forms of capital were needed, both across the field and in the way they were deployed.

Selection and Recruitment of Workers

In terms of the themes of this book, the central issue of the interviews was to gain information about the way the managers solved their human resource challenges. A key challenge for all the managers interviewed was to find workers who had both the technical and social skills needed to be effective in their organizations. The diversity of the field meant that a range of skill sets was required. For example, informants reported that those working at the creative end of the field needed to have good ideas and to be able to communicate these ideas effectively to others; possessing creative ideas was not a prerequisite for many other positions, such as for crew members. Instead, such workers needed to conduct themselves in a military style. In other words, they needed to be highly disciplined, able to follow instructions, and work as part of a team. Indeed, it is reasonable to argue that particular types of workers competed in different ways throughout the field.

Nevertheless, across the field, as one interviewee put it, whatever the position (even those that were advertised), the identification of suitable workers involved "a lot of discussion about who was good and who was not, who is about and who was not." This discussion took place between actors who participated in networks. Thus, as previous research has highlighted, network recruitment was important in the field (de Bruin and Dupuis 2004). Personal recommendations and reputation were important to gaining and retaining insider statuses in all areas of the field, but particularly for those competing for short-term contract work, for example, positions as crew.

Tom: Sally might send her CV in, and she might have worked with John and Claire down the hall, might stick her hand up and say, "look, I'm looking for crew—I'm looking for a really good production secretary," and John will say, "look, Sally's just sent in her CV, and I worked with her two or three projects ago, and she was really good. So give her a call and see if she's available."

Interviewer: So is reputation important?

Tom: It's very important—and what you've done. Have you got the experience? Have you worked on a documentary before? Have you worked as a production secretary before? Have you worked on a big production before? Where was it? Who was it for? How long ago was it?

Interviewer: How would a person enter and progress in the field?

Kim: A good amount of it is it's not what you know it's who you know. Some people get in because their father or mother or whoever knows someone who is going to work on a movie or whatever and they need to employ a runner to make coffee or be somebody's assistant. They get in on that one. Then some of those involved in the movie go on to another in a couple of months and they remember that there was this kid was good. I've seen plenty of people come up through this way.

Although reputation was critical to employment, as the managing director of a smaller production company described, circumstances of time and place were also of significance in securing employment. Here the data resonate with Kettley's (2007) observations that were made in relation to opportunities to learn. Kettley argued that experience that "constitutes a barrier to participation for an individual or a social group, in a given context, at a given time, may in an alternative context constitute a bridge" (Kettley 2007, 344). Although Kettley is referring to participation in higher education, his comments about the significance of time and place in influencing the competition for advancement hold relevance for those wanting to advance in the screen production field. In this respect, the data show that the creative ecologies that continue to evolve in the field were characterized by a complex mix of factors including skills, knowledge, networks, sources of finance, and circumstances of time and place. For example, securing the support of investors for a film required a creative idea for a story, the technical expertise needed to complete the work, and credibility with investors (expressed as saleability). In the following excerpt, the managing director of a film company provides an illustration of this mix in his description of how he recruited a director for a feature film.

> Interviewer: So how did you recruit the director?
>
> Tony: She just walked in the door. We have a longstanding relationship with Stardust Films. It was the company that produced [a major New Zealand film]. . . . Stardust films has an office near ours. There's a symbiotic relationship between our companies. Well, this young up-and-coming director walked into [Stardust's] office and said she'd just read this amazing novel [and asked] "is it possible to get the rights to it?" And they told her we had the rights. . . . [Stardust's managing director] knew we had the rights because of our relationship. . . . Now we're moving ahead with it because the director is saleable—well this director is this kind I needed to make it work for the investors.
>
> Interviewer: What does it take to be saleable?
>
> Tony: Well it depends upon the project. You've got to be someone they want to invest in. . . . If you've had a pile of [film] failures, then no one is going to want to invest in you. But if you've had a success, you're new, up and coming, and the world is looking at you, they will look at you and be saying, "we're keen to see what that young person does next"—there's an air of anticipation. For some projects you need that and for others you don't. . . . So we knew there were people out there who were keen on her—she was marketable. We also knew the [New Zealand] Film Commission were keen on her because she's up and coming talent. . . . It's a very small world. But, actually, it's not just a small world in New Zealand. Internationally, the world is small. We're a microcosm of the wider world. The world all operates like New Zealand. All it is, is knowing people, knowing connections, gaining market intelligence.

However, although network recruitment was often solely used to recruit core creative staff and contractors, some managers also reported advertising some positions in the open labor market. For example, in the administrative side of the businesses, advertisement and interview were typical recruitment methods. Similarly, agencies that knew the company's needs sometimes used to find people to fill some vacancies. However, even in instances where formal processes were used, network resources were important in the process of recruitment, with a combination of open advertisement and the bush tele-graph usually meaning that a flood of applications came in.

Jenny: For a staff position, say, in development [of ideas for screen pro-ductions], we'll get their CVs, and the chances are we will know them already, or at least have heard of them, so they will come with some sort of reputation . . . but then they may not. Then we look at what they have done. We see a lot of CVs, so we are able to judge reasonably quickly, or make an assessment . . . based on the credits they put up, the companies they have worked for, the positions they've held, as to what sort of actual practical work experience they have had. Of course their reputation is impor-tant . . . we'll draw up a short list . . . then we interview.

As noted above, the precise mix of skills and work practices needed by employ-ers varies somewhat across genres and between projects within these genres. In feature film production, for example, workers are expected to stay on set for extended periods of time and to work very closely with their coworkers. These expectations were seen to require social skills that were not as impor-tant in other settings, such as the production of documentaries. Often these social skills were seen to be more important than the technical skills, with managers stressing the importance of social skills such as teamwork. As one informant put it, they preferred to "pick the person who could actually sur-vive with the team, because this would give a better result in the end." Others expressed similar sentiment.

Bob: Personality fit is very important to us. We have very low staff turnover. It is important that the people who come and work here are going to fit in with the group. It is a very busy indus-try. It can be very stressful. There's a lot of pressure. There's a lot of time constraints. You're always working to deadlines. . . . [I]t is important that you have people in the building that are able to work independently and collaboratively. . . . [S]o fit is important.

The desire to recruit those who held the required social skills raises questions about how competency is signaled in the screen production field in Welling-ton. Assessing whether or not candidates had the necessary social skills was not possible through bureaucratic methods, as qualifications did not effectively signal these. Instead, social methods of recruitment were relied on. Man-agers reported that gauging an individual's 'fit' was not a scientific process.

Rather, they relied on gut feelings, their prior experience working with potential recruits, and of course, the candidate's reputation in the industry. In the following excerpt from the data, an employer describes how newcomers to the field can gain insider status:

Interviewer:	So what signals of competency do you rely upon?
Sally:	Can they do the job!
Interviewer:	How do you assess this?
Sally:	You just watch them. You talk to someone else who has worked with them. . . . The key thing is to get [an employer] to give you a chance. You may get that chance, but you are going to work for nothing in your first job, and what you have is someone who will say you did a good job—you have a track record. So person A will ring up person B and say, "I hear you had young John on your job last month, how did he go?" If they say "useless," then you're out. If they say "good," then you're got a chance. If they say "no he's fine, he's good, got a good attitude, got on well with the crew,". . . he's in. . . . A good attitude goes a long way.

In this respect, one manager went as far as to describe the act of making a film as a military exercise, requiring strict adherence to planning and strong discipline.

Sam:	Making a film is a military exercise—requiring strict planning—you are realizing the director's vision. Being lazy, being irritating are not acceptable. You are following directions because time is money. . . . [A]nd there is no such thing as being sick. When you're needed on set, you go.

When considering these comments, it is worth noting that the precise mix of aesthetic skills needed to be effective varies across the field. For example, the skills needed to be effective in positions that work directly with clients differed from those needed in other areas of the screen production field. For example, where workers were required to work creatively with clients, managers reported being very careful to ensure that they employed people who they felt would work well with their clients and not risk undermining the firm's reputation.

Interviewer:	By reputation what do you mean by that?
Kim:	You'd know that they are some one who the clients respect and they would feel comfortable working with. . . . This relates

specifically to short-form, commercial productions; it's different in long form productions. But more and more our clients are very young. . . . They are twenty-upwards-type people and in very senior positions in some cases. . . . So the agencies [who commission advertisements] look to places like ours to make their work as attractive as possible for their clients, so they can keep their jobs. So [the agencies] are looking for people in a place like this who have a reputation. So basically they are purchasing a guarantee that they are going to get a top-class job. If you suddenly try and bring someone in who is not known locally and try and to introduce them and put them on that work, unless they have really got a fabulous personality and the skills to match, you really only get one chance. If they don't gel with the clients first time, you [are] in danger of losing the project.

Finally, another aspect of the ecologies present in the screen production field in Wellington was the way managers worked to create opportunities for creative workers they were keen to recruit—if managers knew of creative workers who they were keen to recruit, they would create positions for them.

Sally: Historically, for producers, we have come across people who have maybe come back to New Zealand. . . . If we think the person can bring value to our company, we'd create a position to accommodate the person's skill set—we can benefit from their skill set. So the employment of creative people can be kind of an organic process.

Sources of Innovative Knowledge

A central aim of this book is to explore how firms access knowledge that provides them with competitive advantage. In the screen production field, innovative knowledge is understood to be that which could be used to produce and develop content that has commercial value. As shown in the previous section, recruitment through social networks provided a key source of knowledge that had this potential. It is also important to point out that the ideas for creative production often come from brainstorming ideas of employees within the firms. In instances in which the initial idea for a production came from outside the firms, story concepts were further developed through collaboration. Thus, the process of being creative was often, though not exclusively, a collaborative exercise.

However, this only tells part of the story. Knowledge about what projects would meet with positive responses from viewers was also required. According to the managers, this is not a skill that managers reported could be taught in a formal way, for example, in institutions of tertiary education; it is partly an aesthetic skill, which includes having knowledge of consumer preferences. This knowledge was perceived to only be gathered through social interaction with the funding agencies and other purchasers of creative context, and, thus, access to it remains closed to outsiders.

Sally: I'm dealing on a weekly basis with the commissioners for the major networks, so we know what programs they are looking for. . . . It's not knowledge that's held in the public domain. . . . We try and go higher in the organizations [that purchase creative content], up to those who make the policy decisions. Even though these people do not tick things off, they are the ones who ultimately hold power, and these people are the ones who can tell you what the networks are really looking for. And it changes from month to month, so this relationship needs to be ongoing. . . . Go back a step. If you have an idea for a documentary, how the hell do you get the documentary made? Well, first of all you have to go to the television stations, and you have got to form a relationship with them and get your idea accepted—but they do not know who the hell you are. So that's a factor. However, if they like the idea, they will like it. They will do it for the gold. But then you have to understand the process. They have got to be confident that (a) you can deliver it—could you make it? Then (b) you have got to find the money, and this means, in the case of documentaries, that you might need to approach New Zealand on Air. So then New Zealand on Air have to know who you are—to be confident about you and like the idea. They don't automatically fund projects. You have got to convince New Zealand on Air that you can do it. So there's a whole pile of relationships in there that are really critical, and that's what a company like ours brings to the party—we have those relationships already. . . . I initially built the relationships and have transferred them to others in my company. Don't underestimate the ideas—they are really important—but you can have the best idea in the world—one that everyone thinks is crash-hot—but if New Zealand On Air don't trust you, if they don't trust you to manage it, or to see the project through to fruition, even if it's the best idea in the world, then they are not going to go with it.

Thus, these market makers traded on their social capital to gain intelligence about what the networks and others were interested in funding. In other instances, however, star film makers such as Peter Jackson had been so successful that they were able to partially finance the productions themselves.

Links with Providers of Tertiary Education

As outlined above, growth in film has been matched by growth in the provision of a diverse range of training. Although the Film and Electronic Media Industry training organization folded, training linked to the New Zealand Qualifications Framework is available in most, if not all, areas of screen production. As part of assessing the way in which they build and use networks, managers were asked about their relationship with providers of tertiary education. At a general level, the data show that the relationship between tertiary education and advancement in the film field is, at best, weak. Producers and directors have often been self-taught (with Peter Jackson providing a prominent example), and there did not seem to be any connections between the production of innovative or creative ideas and training in the tertiary education sector. When asked whether they preferred new recruits to have formal training and qualifications of any kind, the following responses were typical.

Sally: [Formal training] is not going to help you much. It may give you some craft knowledge, but the truth is that when you leave school, you've got to start again. You might think you know the business, but you are not going to know how it works in Queenstown. . . . Partly it's to do with the quality of the film schools. Partly it's to do with the nature of the industry—we've all had to start at the bottom. . . . I think it's a combination of both. . . . The company has taken a couple of kids on from a couple of places because we know that they have been given realistic expectations. . . . It's not like getting a teaching diploma, which qualifies you to teach. You get your qualification in film, then you start at the bottom. . . . There's no guaranteed route into the industry at all. Because it's tied up with personality and personal initiative and creativity. . . . Graduates start in the most menial of positions. . . . They're going to be runners—gofers.

We have recruited from the film schools, but only because they are a convenient source of labor. . . . The value added by the school might be [little], it's just a convenient place to go [for workers].

Interviewer: Do you recruit through some of the film schools?

Bob: It depends on the requirements of the position. For the position we have advertised at the moment, the chances of us employing a film school grad depend on what they've done before their being a film school grad. A lot of film school grads don't go to film school straight from secondary school. In fact, they do five different things before they work out [that] what they actually

Interviewer: want to do is go to film school. . . . So they could have all sorts
of really interesting and useful experience prior to having gone
to film school and come out with some understanding of what
the film process is all about. In this case, they would very defi-
nitely be considered for the position. But if we are talking about
someone that has pretty much gone straight from school, maybe
a year or two travelling about, the chances of them getting
employed are pretty slim, because they just do not have the depth
of work experience.

Interviewer: Where do people like that go to get a job?

Bob: I have no idea—a lot of them wait at tables. It largely depends
where they want to go. If we're talking crew, they start as runners . . . sweeping the floors, making the coffee, running about,
working 14 hours per day, trying to get to know everybody in
the industry that they can possibly get to know so they can
get recommended on the next job. That's what happens to
those film students. Those who want to work in production
offices go from North Cape to Bluff trying to get a job in a
production company.

One reason such testimony is interesting is that the managers specifically
identify that the rules of advancement differ in their field. Innovative knowledge is not transferred from knowledge generators into providers of content
in any detectible or direct way. Similarly, there is little emphasis placed on
formal tertiary qualifications as a means of determining which workers have
the required skills, and it is therefore reasonable to conclude that tertiary
education is poorly linked to employment in the field. However, one manager did report some positive linkages with those providers of tertiary education that they perceived to provide quality graduates. Thus, there exists a
very small amount of reputational capital, but circumstances of time and
place were also critical. Again, however, it is important to stress that new
recruits who came from this provider start off at the bottom and are required
to prove themselves.

Interviewer: So tell me about the internships you have offered.

Bob: The Christchurch Polytechnic Film School is the one film school
that we will consistently look at the interns because the calibre
of the interns is quite high, and it is a very good degree, and the
syllabus seems to cover a lot of good material, and their tutors are
obviously quite good. We have taken them on production positions. . . . You pay them five eighths of not very much, which for
a struggling production house is a bonus. You work with them
and they work with you. . . . One case worked out very well, and
an intern worked with us for three years, before moving overseas.

. . . We select the interns . . . and will only take them on if I think there is a reasonable chance of them getting a job at the end of the internship. . . . Historically, we have had a good track record with students we have taken on from there, so anyone who comes out of the Christchurch Broadcasting School, you go "umm interesting." And in fact our [production] manager, who has been with us from about five years, has come through the broadcasting school, so again there's a reputation there that a pretty high level of student comes out of there.

Kim: We have taken on a few people who have come out of tertiary training, particularly the Christchurch Broadcasting School. We have two guys who have come here . . . as graduates of the school, who came here on understanding that they would be runners making coffee and on the understanding that you actually had to be good at that job before you'd be considered for any this else. . . . Of course, they did do a good job, and we saw that they had skills—probably even more skills than would have expected from someone who had just done an academic course. And so we moved them on.

Interviewer: Tell me about how they were appointed?

Kim: They graduated, or were near to graduating [from the Christchurch Broadcasting School], and they wrote saying they were prepared to relocate and were happy to come in [to the firm] under any capacity.

Interviewer: So were they cold-calls effectively?

Kim: Yes . . . the only time I've advertised for a position in the last couple of years is for an admin-bookings person.

Interviewer: So was there anything about the applicants, these cold calls . . . from graduates of the broadcasting school that enabled you to offer them a position that would not have gone to another person.

Kim: A lot of it was the circumstances of the time. Runners, as we call them, are generally short-term positions because no one wants to make a career of making coffee and client hospitality, which is what the job is. So you know that you are going to at least have to replace one or two a year. If they are any good, they will want move on to something else and make more of a career out of it. They just happened to apply at a time when we had a couple of vacancies. That plus the fact that they were enthusiastic and they had demonstrated an interest in the industry, put them above other people who may, for example, have been working in a dress shop, but wanted to get into television.

However, if the links with the providers of private training and the polytechnics were tenuous, then the links between the universities and the film field were nonexistent. As one managing director put it, the universities are more concerned about film as art. Apparently, in the university setting, the subject of film has little to do with the application of filmmaking. As one informant put it, "We actually do it, they [the universities] talk about it. We are driven by commercial concerns."

Finally, in instances where managers were looking for a particular set of skills for important roles in their firms, managers reported that they would hunt for the best person possible. This involved digging in to their social networks. In part, this was because there were plenty of people who held the formal qualifications required to be effective in the positions on offer. However, it was also because the real qualities they were interested in required them to use other recruitment strategies.

Jasmine: If I'm looking for a researcher, for example, if I'm looking for any person with established qualifications, I have endless people coming in the door who have many qualifications, who want those positions. There's no shortage of people out there, really. If I want a person for one of these kinds of positions, even though there might be a lot of people banging on my door, I am going to hunt for the person I know can work with the director, who can work with the team—I am going to want the right person for the position. So I am going to hunt for the right person for the position. We know who's out there. In our country, there are not many out there we don't know about, and we can soon find out about these people. We go hunting—we ring them up and ask: "when is your [current] job finishing?" We don't advertise.

Conclusion

The purpose of this chapter has been to explore the human resource practices of managers working in the screen production field in one center and to use this to demonstrate competition for advancement through tertiary education. As discussed in Chapter 2, exploring competitive processes within one field is useful because the rules of advancement are likely to be field specific. The aim of the chapter has been to use the managers' accounts of their human resource strategies to shed light on debates about the competition for advancement through tertiary education.

In general, the data support social network theory, showing that the managers use and make networks at most stages of production and across genres.

A critical aspect of this is the use of networks as a way of gathering knowledge about the potential of new recruits. Many of the skills needed to be effective in the field are not signaled by educational qualifications. Indeed, it seems that much of the creativity needed to work on the creative side of the field is not taught by the providers of tertiary education that appear to directly service the screen production field. On the other side of the field, where workers are required to be highly disciplined and work like soldiers, networks are also vital to advancement.

This is not to say that tertiary education does not support the production of workers in the field in other ways. Rather, it is to say that managers do not look for particular kinds of qualifications, nor to particular institutions to supply their human resource needs. For creative workers, it is the quality of their ideas and their attractiveness to buyers of content that counts. Knowledge about who had creative ideas (or innovative knowledge) and the attractiveness of these to investors and to buyers of creative content was only available to insiders, or those who participated in the right networks.

The lack of apparent links between participation in tertiary education and employment in the creative field raises questions about processes of social reproduction. It could be argued that participation in training linked to the screen production field does not provide a conduit into the screen production labor market. Indeed, it may arguably be a trap for some. Thus, participation in training linked to the screen production field creates what could be euphemistically called a dead end for those who either do not understand the rules of advancement in operation or lack either the creative skills or the contacts needed for advancement. The apparent acceptance of the training gospel suggests that these students are not well embedded in the social networks of a kind that could have provided them with quality advice about the rules of advancement in the field.

Although field theory provides a way to isolate the precise resources needed for advancement and allows for the easier identification of the rules of advancement in operation, defining the field in the manner outlined earlier has limitations. In this respect, the data show that the screen production field is more diverse than implied by the approach adopted in this book, with managers reporting the use of different strategies for recruiting staff in different contexts and across time and space. Although there may be field-specific knowledge that gives advantage, some skills needed to be effective are arguably sufficiently generic to be of use in other fields. Administrators, for example, tend to compete in a discrete field.

CHAPTER 6

Innovation and Networks in the Knowledge Economy

The Case of the Biotechnology Field in New Zealand

Introduction

As noted earlier, the New Zealand Labour-led Coalition Government of 2005–2008 sees itself playing a key role in helping to develop research and development infrastructures that promote innovation. As the Minister for Economic Development and Industry and for Regional Development put it recently: "Innovation can only occur when the institutional and organisational frameworks, regulatory systems, infrastructure and processes for diffusion of technology are mutually supportive. . . . The economic transformation themes are similarly interdependent. In order to grow more globally competitive firms we need innovative and productive workplaces underpinned by world-class infrastructure, a sustainable natural environment, and an internationally competitive hub in the form of Auckland. Links between industry and science are critical as many forms of innovation require a combination of market knowledge and technical know-how, which can't be achieved by a business or research organisation working alone" (Mallard 2006).

To support the development of innovation, the Labour-led Coalition has set about reforming tertiary education in New Zealand. A central aim ofthese reforms is to concentrate research capacity in relatively few institutions and to build networks among creators of innovative knowledge, employers, and financiers.

This chapter explores the biotechnology field in Auckland, New Zealand, in relation to network formation. The purpose of the chapter is to use the testimony of managers to assess how the labor market for skill and innovative knowledge functions in the field of biotechnology in New Zealand. In exploring the behavior of managers, it considers the role of tertiary education in mediating the competition for advancement in the field of biotechnology. The field of biotechnology has been selected for this case study because, as with the creative field (see Chapter 5), it is held up as evidence of New Zealand's ability to compete in the global economy in ways that promote growth in knowledge work, and also because it is an area in which the state has claimed that its interventions have made an important, if not critical, contribution to growth in knowledge work.

The second section of the chapter briefly describes how government is attempting to build knowledge ecologies in biotechnology through network formation and related reform of tertiary education. Next, the chapter draws on reviews of the field (L.E.K. Consulting 2006; Marsh 2004) to provide some background information on the biotechnology field, and then describes the method used to gather the data on which the chapter is based. The following sections draw on interview data with managers.

Building Knowledge Ecologies in Biotechnology New Zealand

In recent years, considerable attention has been focussed on New Zealand's biotechnology sector. Politicians, policy makers, and commentators have come to see biotechnology as offering an opportunity to grow the economy in ways that increase demand for knowledge workers in environmentally friendly ways. Indeed, as noted in previous chapters, the Growing an Innovative New Zealand Framework (and the Economic Transformation Agenda that followed) specifically identifies biotechnology as one of three areas of the economy worthy of receiving special attention and intervention by government (Office of the Prime Minister 2002). In 2002, the then-government published the Biotechnology Taskforce's Report. The task force had been established to identify ways to promote growth in the sector. This led to the publication of the New Zealand Biotechnology Strategy (Ministry of Research Science and Technology 2002), which identified three goals supportive of the government's vision: These were:

1. Build understanding about biotechnology and constructive engagement between people in the community and the biotechnology sector;
2. Grow New Zealand's biotechnology sector to enhance economic and community benefits; and
3. Manage the development and introduction of new biotechnologies with a regulatory system that provides robust safeguards and allows innovation.

As shown throughout this book, reform of the tertiary education sector in New Zealand, which is underpinned by recent and current governments' desire to direct the provision of tertiary education, is an important part of this strategy. The idea is that deliberate intervention by government can lift innovation.

The extent to which governments can successfully identify and nurture emerging areas in an economy has been widely debated (for an outline of the arguments, see Krugman 1986). Nevertheless, even though the debate about the governments' possible contribution to the process remains unresolved, there is strong empirical and theoretical support in the literature for the idea that innovation is encouraged in biotechnology by the presence of linkages between firms that encourage knowledge spill-over. For example, Arora and Gambardella (1990) conclude on the basis of their analysis of external linkages of large chemical and pharmaceutical companies in the United States, Europe, and Japan that linkages are positively related to innovation (after controlling for firm characteristics). On the basis of this finding, they argue that small firms are now an important source of innovative activity and that the locus of innovation should be thought of as "a 'network' of inter-organizational relations" (Arora and Gambardella 1990, 374). In the New Zealand case, Marsh's (2006) research produced similar results. He found that smaller biotechnology firms were more innovative than large firms and suggests this may be a result of their ability to share knowledge.

A further branch of research explores whether or not networks identified by researchers such as Arora and Gambardella (1990) are likely to be international or intranational in nature. On the one hand, Andersen (1992), for example, stresses the importance of intranational linkages on the grounds that they are a better means of transferring semiformal and informal information. In addition, the formation of such linkages is easier within nations. One reason for this is that the knowledge needed to innovate can only travel for short distances.

On the other hand, more recently, Johnson (2002) has challenged this view, arguing that in biotechnology, developments in technology mean that geographical barriers no longer present the barrier to knowledge spillover that they once did. Because of this, it is no longer as important for biotechnology firms to cluster in discrete areas. In related research, Keller (2002) concludes that developments in national technology have often ceased to play the most important role in the nine Organisation for Economic Cooperation and Development countries he studied. Instead, he found a strong tendency toward the globalization of technology and identified language skills as being of importance to knowledge spill-over. Similarly, Marsh (2006) argues that international linkages may be of increased significance for small, geographically isolated countries such as New Zealand. Indeed, Johnson (2002) suggests that in the contemporary period, biotechnology is more likely to jump long distances than has traditionally been the case, raising the possibility of the deliberate fostering of biotechnology in areas of the world where it has not been present to date. A prerequisite for such movement is the presence of vibrant communications linking together distant research. The location of these communities is strongly influenced by the location of super-stars in academic research. In other words, biotechnology tends to be located in close proximity to leading academic researchers (Zucker et al. 1998).

As noted above, the Labour-led Coalition has accepted that state inter-vention can contribute to the development of innovation. It argues that to date, neoliberal methods of funding have encouraged providers of tertiary education to focus their attention too heavily on increasing the numbers of students participating in tertiary education and too little on the quality and the relevance of the training offered. As a result, the quality of the training offered has been poor, limiting the contribution made by the tertiary educa-tion system to New Zealand's innovative capacity. To rectify the situation, the current government has settled on two main strategies. The first is the introduction of the Performance-Based Research Fund, which is designed to concentrate research capacity by directing funding for research to those institutions that conduct research. The second is encouraging network for-mation in the form of collaborations among knowledge generators, private businesses, and sources of venture capital. These two, interrelated, strategies are considered in turn.

Concentrating Capacity

As noted earlier, a basic premise held by the Labour-led government is that the free-market policies advanced by previous administrations in the areas of research and development and in tertiary education have not adequately promoted innovation or delivered training that is of sufficiently high qual-ity. A key reason for this advance by the current government is that funding was spread too thinly across all providers of tertiary education. This reduced capacity by diluting the state's investment in research and development. According to government, rectifying this requires that capacity be concen-trated in relatively few institutions (possibly to support the work of relatively few "stars").

A key policy to concentrate capacity is the Performance-Based Research Fund measure. The fund is designed to help lift capacity to innovate by more generously funding those tertiary institutions that are conducting research. As the then–Minister for Tertiary Education expressed it, the

Performance-Based Research Fund is designed to allow funds to "follow demonstrated research performance, rather than being spread thinly across all Tertiary Education Organizations irrespective of their research output" (Maharey 2004). As noted earlier, in the 2003 round, the University of Auckland topped the ranks in terms of gaining sources of external research income, degree completions (Masters and Doctorate levels), and in terms of quality ratings (primarily, the quality of peer-reviewed publications). As a result of their superior performance in this and other areas related to research, the University of Auckland received nearly one third of the Performance-Based Research Fund in the first round. The University of Otago came in a distant second place. Combined, Auckland and Otago Universities earned well over half of all external research income available in the sector in that year.

In the 2006 assessment round, the University of Otago lifted its performance markedly, but depending on how one views the data, its performance remains below that of the University of Auckland. When this success is translated into financial terms, Auckland and Otago Universities will receive the bulk of the Performance-Based Research Fund from 2008 until the outcome of the 2012 round is implemented in 2013.

At the other end of the scale, following the first round, many institutions saw their funding stagnate or even decline as a result of the Performance-Based Research Fund. For example, the Open Polytechnic saw its funding for research decline by almost 50 percent, and the now-defunct colleges of education also lost funding. The 2006 round[1] confirms this trend (although the shifts were less dramatic), with less money going to research-inactive institutions.

Another way government is concentrating capacity in relatively few institutions has been in moving away from funding tertiary education organizations simply according to the number of students enrolled and toward funding them according to a complex formula, which recognizes the institution's research and teaching profiles. As part of a general move to increase the role of government in determining valuable knowledge, the Tertiary Education Commission has required that all tertiary education organizations develop profiles of the kinds of research and teaching they will undertake. Profiles will be negotiated between the Tertiary Education Commission and individual tertiary education organizations. The intention is that government will negotiate with providers for the delivery of courses it wants, so that a network of provision is developed. This move is designed to drive up the quality and relevance of research and teaching in the sector by reducing competition and duplication of provision. As part of the general strategy to drive up the relevance of the training offered, the Tertiary Education Commission is also requiring that the institutes of technology (or polytechnics) report on the skill needs in the regions they service and demonstrate how their courses satisfy these.

Network Formation

Network formation is a second and related strategy employed by the current government to boost the capacity to innovate. Here, government is encouraging tertiary education organizations to form cross-border collaborations, particularly with industry (Howells and Nedeva 2003). As has been widely canvassed in the literature, the idea is that private sector funding should be sought to subsidize the work of universities. In general terms, the formation of networks by the government reflects a growing awareness that creating markets in education and training of the kind that drive up innovative capacity requires the creation of networks between knowledge generators and innovative firms. Moreover, the concentration of government and industry funding for research in relatively few institutions arguably makes economic sense because it has increased the ability of government to invest more heavily in areas of strategic importance.

Links between the business sector and the tertiary education sector are not new. For example, there exists a long history of universities forming collaborations with private businesses. As Bok (2003) explains, greater collaboration is, in part, a result of reduced funding by the state for higher education, which has forced universities to seek external sources of funding. In addition, universities themselves have been keen to expand their domains and to meet the expectations of academics for increased remuneration and for greater support of their research. As Bok (2003, 9) points out, universities "share one characteristic with compulsive gamblers and exiled royalty: there is never enough money to satisfy their demand." Accessing sources of financial support from the private sector is good way of meeting such challenges. Yet another view is that private enterprise wants to commodify higher education and perceives providing financial support to be a good way of achieving this. In this respect, the commercialization of higher education in the United States was more prevalent after 1980 because the financial rewards gained from providing expert technological skill, expert advice, and expert knowledge were increasing (Bok 2003). Until the latter part of the twentieth century, the commercial value of many discoveries made by academics was not readily apparent. At the same time, the business community did not generally look to universities to increase their competitive advantage. In any case, academics were not generally motivated by commercial imperatives and tended to see their work as contributing to the public good. However, in the more recent period, recognition by governments that providers of innovative knowledge can enhance competitive advantage, and increased awareness and desire on the part of researchers to commercialize their discoveries, has seen an unprecedented drive to promote links between business and tertiary education and to commercialize intellectual property. In this respect, Bollier (2002) remarks that legislative changes have enclosed the academic commons, reducing the extent to which innovative knowledge is democratized.[2] As he points out, until recently, much scientific work was undertaken as a contribution to what is sometimes referred to as the gift economy. However, the commercialization of tertiary education has undermined this, and the rules that typically govern the use of any innovation created through joint research projects financed by business may limit the dissemination of findings to others. Other researchers suggest that rather than enclosing the academic commons, commercialization need not limit access to research findings, and strategies such as defensive publishing may actually help by increasing exposure to innovative ideas (Adams and Henson-Apollonio 2002). By publishing their findings, any innovation becomes seen as "prior art." This prevents competitors from gaining exclusive patent rights to innovations produced and effectively keeps the owner's patent as the only exclusive portion of an invention. In turn, other firms may see the potential that the innovation might hold for their own innovations and seek out partnerships with the developers of the initial innovation.

In New Zealand, the government has used a mixture of measures to promote links or networks between tertiary education organizations and business. One measure of encouraging network creation is simply to force tertiary education organizations to seek other sources of funding (by, e.g., reducing the contribution made by government).

Government is also encouraging greater collaboration between businesses and tertiary education organizations through more direct methods, such as funding network formation directly. One such direct intervention has been to fund seven Centres of Research Excellence. These were established in 2002–2003 to encourage the development of world-class research in New Zealand and to facilitate knowledge-transfer activities. The stated aim "is to support leading edge, international standard innovative research that fosters excellence and contributes both to New Zealand's national goals and to knowledge transfer" (Tertiary Education Commission 2004, 2). To be considered for funding, providers need to demonstrate how their collaborations will contribute to New Zealand's capacity to innovate and how the knowledge produced will be commercialized.

An important aspect of the Centres of Research Excellence is the formation of interinstitutional research networks. Each center is hosted by a university and includes a number of partner organizations. These may include other universities, Crown Research Institutes, and Maori providers of tertiary education (or wananga). Among the host institution's responsibilities is support for knowledge transfer and network activities.

Thus far, the government has funded seven centers, within which a critical mass of researchers, private companies, and other resources are pooled to create economies of scale, reduce transaction costs, and enhance the development of commercial products. The University of Auckland hosts four of these centers, including New Zealand's premier biotech research center, the Maurice Wilkins Centre for Molecular Biodiscovery, which has partners in other centers.[3]

Network formation is also being encouraged in other areas of the tertiary sector. For example, the creation of the Institutes of Technology and Polytechnics Business Links Fund is designed to foster greater engagement among institutes of technology and polytechnics and business.

Other measures include the Partnerships for Excellence framework that aims to ensure that the tertiary education sector develops so as to meet the needs of the private sector and industry. By increasing the level of private sector investment in tertiary education and by fostering better links among tertiary education organizations, industry, and business, it aims to make tertiary education organizations more responsive to New Zealand's social and economic development needs. Partnerships for Excellence enables tertiary education organizations to seek matched funding (either as a capital injection or as a suspensory loan from government) for large-scale investment projects that are unable to be funded via other means (in general, those valued at NZ$10 million or more). In 2004, the University of Auckland won funding of NZ$10 million to support the Institute of Innovation and Biotechnology to provide a center for graduate training and research in biotechnology. According to the Tertiary Education Commission (2004), at the institute, "graduate students are immersed in an entrepreneurial environment where research and its translation into commercial applications are co-located. The Institute for Innovation in Biotechnology is equipped with state-of-the-art technology and works closely with the commercial sector" (2004, 2). These and other measures mean that universities now rely much less heavily on government funding. However, as noted throughout this book, the level of funding provided by nongovernment sources varies considerably from provider to provider within the tertiary sector.

Background to the Field

In New Zealand, biotechnology builds on national strengths in agriculture (particularly its 150 years of experience in improving animals and plants), but there are a number of important developments in areas such as health that have occurred in the more recent period. Other researchers have noted that different interpretations of the terms biotechnology and biotechnology sector have hindered attempts to measure activity in biotechnology across time and space (Marsh 2004). Bearing in mind this caveat, Marsh (2006) suggests that the field in New Zealand is characterized by a wide variety of firms. Indeed, the number of firms ranges from approximately 30 core biotech companies with a combined annual income on the order of NZ$200 million to many thousands of companies having annual production worth several billion dollars (when traditional food products such as cheese, yoghurt, and beer are included in the figures). Taking a wider view of the field, since 1999, the sector has experienced strong growth, employing 2,200 people in 126 organizations and netting expenditures of more than NZ$640 million in 2004. Between 2004 and 2005, expenditure on biotechnology increased by 20 percent, to reach NZ$640 million (L.E.K. Consulting 2006).

In terms of expenditure, in the year 2004–2005, approximately 40 percent of activity (as measured by average expenditure and employment) was in agricultural biotechnology. Medical and diagnostics accounted for 23 percent of expenditure and 28 percent of employment, with industrial applications accounting for 19 percent and 14 percent, respectively, and human health applications for 17 percent and 23 percent, respectively, making up the rest of the expenditure and employment activity.

The government's share of expenditure on research and development has been estimated to be approximately NZ$127 million per year (Ogilvie, cited in Marsh 2006) and makes up around 19 percent of total government research and development spending. This makes New Zealand one of three Organisation for Economic Cooperation and Development countries with a share above 10 percent (Organisation for Economic Cooperation and Development 1996). There has also been significant investment in research and development by private companies such as Genesis, which has technical expertise in the fields of genomics and immunology, and which has invested over NZ$80 million in research since its inception in 1994. Although the level of investment is high by New Zealand standards, Marsh (2006) argues that New Zealand's total expenditure on biotechnology research is very small by global standards.

Consistent with Marsh's (2006) suggestion that New Zealand's geographical isolation means it would be more likely to rely on international sources of knowledge, it seems that in New Zealand, biotechnology international collaborations are on the increase. Indeed, in the years 2004 and 2005, approximately 650 partnerships were formed with a wide range of public and private sector organizations. Seventy-five percent of these were with international partners, particularly from Australia (L.E.K. Consulting 2006). Although the nature of these partnerships remains unclear, it is likely that they primarily involve sharing intellectual property and other forms of expertise.

It is also important to note that activity in biotechnology is regional in nature, with the Auckland region accounting for the greatest proportion of income (35 percent of national income) and expenditure (30 percent of national expenditure) in 2005. The Waikato and Central regions were also significant contributors to biotechnology activity (L.E.K. Consulting 2006).

Growth in the field has been matched by growth in the number of people leaving the tertiary sector with degree-level qualifications in biological science. Between 1981 and 2001, the number of people who reported degree-level qualifications in this area increased from 4,584 to 12,036 (New Zealand Ministry of Education 2006). Although figures for the number of people graduating with postgraduate level qualifications are not available, the increase in the production of graduates at the undergraduate level is likely to be mirrored at a postgraduate level. For example, L.E.K. Consulting (2006) estimates an increase in the number of postgraduate degrees in biological science and in health from 500 in 1994 to 700 in 2004. They also report that postgraduate qualifications were the most common qualifications held by employees in the organizations they surveyed in 2005. They found that 34 percent of employees held PhD-level qualifications in 2005, down from 40 percent in the previous year. They also noted that the greatest increase in employment was for those who held technical and trade-level qualifications. Growth in employment overall increased by 30 percent between 2004 and 2005 in private sector organizations, whereas employment in the public sector and institutions of higher education declined by 6 percent and 2 percent, respectively (L.E.K. Consulting 2006).

Along with growth in domestically trained workers, employers in biotechnology are also increasingly seeking staff from overseas. This most likely reflects both an outcome of growth in the number of international collaborations and general difficulties employers report gaining suitably skilled workers. Whatever the reason, between 2002 and 2005, the number of work visas approved in biotechnology and related areas increased from 99 to 215.

Having outlined the size and scope of biotechnology and its relations to the tertiary education sector and to government policy, attention now turns to the empirical section of this chapter.

Method

The method employed to gather data was similar to that used in the previous chapter. Senior managers working for larger biotechnology firms, midsized enterprises, and smaller start-ups were interviewed as part of this case study. All of the managers were either chief executive officers or chief scientific officers, and all had ongoing involvement with, and knowledge of, the recruitment processes. In addition, recruitment focused on the knowledge creation end of the scale. An aim of the interviews was to map the human resource practices across the field and to use this to shed light on its relationships with the tertiary education sector, particularly as these relate to sources of innovative knowledge. For this reason, emphasis was placed on recruiting participants who worked at the discovery end of the field. An initial list of potential informants was selected from the New Zealand Biotechnology Industry Training Organisation's Web site. All of the informants were based in Auckland. In total, 17 managers were interviewed from across a range of different types of providers. The same method as that employed in the previous chapter was used to gather data and to report back to participants (see Chapter 5). The participants suggested a small number of changes to the chapter. All of these were minor and were incorporated in the final version.

The Biotechnology Field in Auckland

The central research question that this chapter is attempting to answer is: How do biotechnology firms in the Auckland cluster make and use networks to solve their human resource problems? The evidence gathered points to the importance of networks as a means of gaining access to resources of varying kinds at most stages of production. However, as described below, the way networks are used varies considerably across the field and changes as firms evolve.

For example, as biotechnology firms move from the start-up phase through to the production phase, their human resource needs changed. In the start-up phase, the firms often rely on networks to secure suitably skilled labor. In contrast, in the production phase (which relatively few biotechnology firms had entered), or for those firms that focused on producing established products, it was more likely that firms would use advertisements and interviews to recruit staff. This method (advertisement and interviews) tended to be used to recruit workers who could operate the machinery, complete routine testing, and keep the records needed to meet the regulatory requirements of government authorities. Similarly, advertisement and interviews were used to recruit administrators. However, as noted above, the use of these strategies suggests that such workers actually compete in discrete fields that have different rules of advancement.

It would be a mistake to see the firms' evolution as necessarily linear. Rather, conceiving of production in these ways provides a useful heuristic for better understanding the managers' recruitment strategies. However, in general, and consistent with the literature, the interview data gathered as part of this case study indicate that social networks remain critical to advancement in the field.

The key themes that emerged from the data were sources of innovative knowledge, the selection and recruitment of new workers, and links with tertiary providers. These themes are discussed in turn and in relation to different stages in production just noted.

Sources of Innovative Knowledge: The Role of Star Scientists

All of the interviewees reported that the original ideas on which their firms were based came from with the universities and other largely government-funded organizations, such as the Crown Research Institutes. The precise way the ideas were discovered and exploited varied. Nevertheless, all firms could trace the origins of their products back to discoveries made within government institutions. These discoveries were then combined with other resources, such as sources of finance, and unique features of New Zealand to produce commercial activity.

In some instances, circumstances of time and place, combined with personal networks, were significant to producing commercial activity. For example, in one firm, an encounter between a scientist, who had spent much of his life searching for a solution to an enduring medical problem, and a wealthy investor, who participated in the scientist's networks and who had heard of the innovation during a social event, led to the formation of a new biotechnology company. In another firm, a discovery made by a star scientist in another country that could be advanced by accessing generic material only available in New Zealand led to the establishment of a biotechnology firm.

In others, the bringing together of these resources was more deliberate and planned. Nevertheless, these findings remained consistent with the literature, which points to the importance of star scientists and participants in their networks in influencing the geography of biotechnology, particularly in its early stages of development. In most cases, innovative knowledge created in government-funded institutions by individual scientists provided the initial innovation on which the biotechnology firms were based. As the following data attest, once the decision had been taken to progress the innovation, the star scientists set about building research teams.

Tom: The company has taken compounds that were discovered at the Medical School, University of Auckland, that have been developed to solve principally two main medical problems lung cancer and divertulosis. . . . As a result, we have two distinct arms to our enterprise. . . . Both solutions come out of the university, but there is a little bit of technology from overseas.

Interviewer: So someone in the university discovered the technology and saw the commercial potential?

Tom: Yes, this is essentially the model. The university had a major role in promoting the technology. There are other participants. Our Chief Scientific Officer [who continues to work in the university] and one of our lead investors—they all came together and saw the benefits of some progress jointly. I think you can say that all three were drivers of the project. So it comes down to key personalities and leadership at the end of the day. . . . Our Chief Scientific Officer discovered the first two pieces of technology. He was responsible for bringing people into the company and for finding people who were doing interesting things and persuading them to transfer their technology into the new company. . . . What's happening in New Zealand is that the biotechnology companies essentially have developed around world-class scientists who have attracted reasonably world-class people around them.

It was reported also that firms gain access to knowledge through the publications and registered patents of other researchers that provide information about developments in the field and information about whom to contact for more information. The cross fertilization of ideas between academics and researchers in private firms was vital to advancing the science that provided the foundation on which the biotechnology rested. In this respect, there was some feeling expressed by the participants that New Zealand's researchers were not making the most of their innovative ideas. To make the most of these ideas, entrepreneurial skills were needed. Some interviewees described how they had made the shift from being purely researchers in academic disciplines, such as biology, to being entrepreneurs. For example, a manager of a midsized firm (100-plus employees) described how her ability to span boundaries between academia and business was critical to their success.

Mary: I saw an advert for a job as a scientist at Impact Biosciences, as it was known in those days, and I saw that you could do both science and business. And I thought this was interesting because all my [university] students were turning up to class in big, flash cars, and I was driving a Morris Oxford. . . . I figured you could do business and do science. . . . If you could get out into industry, there were a lot of opportunities, and if you could be good at

> the business of sales and marketing, and be good at science and
> technology, you could actually have the benefits of both worlds.

In part, it was the scientist's ability to speak the language of both science and commerce, or to span boundaries between academic disciplines and business, that led to the commercial activity. Indeed, as noted above, this is a defining feature of biotechnology.

It is important to point out that, in the vocabulary of economists, knowledge is really only innovative if its commercial potential is realized. What made knowledge created in the universities innovative was the way in which the formation of teams enabled innovation by facilitating boundary spanning. By combining the resources of different people and sources of finance, innovative ideas were commercialized. In effect, the network of individuals enabled boundary spanning. Thus, boundary spanning can be as much an attribute of a team as it can be an attribute of an individual. In this context, it is interesting to note that MacDiarmid attributed some of his success in the interdisciplinary nature of his research to the collaborations he developed with other researchers (Dekker 2007).

The data reveal, however, that although the universities and other research institutes provided the initial innovation, further development of innovations tended to depend on gaining access to international sources of knowledge. The managers did not employ a single strategy to gain this knowledge. Instead, a range of strategies was used, including recruitment of new staff (both nationally and internationally), using contractors, attending conferences, personal communications with other leaders in the field, and reading the research publications in journals. For some managers, the key to gaining innovation was not holding technical knowledge within one company but, rather, as one manager interviewed as part of this research commented, knowledge about "whom to ask for advice and to know if the advice is sensible or not sensible, and who to devolve decision making to." Nevertheless, a key way the managers gained innovative knowledge was through recruiting new workers who held the requisite skills.

The Selection and Recruitment of Skilled Labor

To understand properly how firms gained and maintained competitiveness through, for example, gaining access to innovative knowledge, it is important to understand their human resource strategies. In this respect, as noted, as firms evolve from start-up and progress through various stages to production, their human resource needs change.

At a very general level, in phase one (or the start-up phase), the tendency was for local discoveries to be combined with sources of finance to develop products and to ascertain whether these had commercial potential. During this phase, the emphasis was on further scientific discovery and testing products. As teams were formed, and as the products were further developed, recruitment through domestic and international networks was critical. One reason identified by participants for the importance of networks is that the skills needed to innovate are highly specialized and that relatively few individuals in the world hold them. Thus, in contrast to Durkhemian-inspired functionalism, in which it is predicted that the use of open recruitment methods increases as production becomes more technocratic and scientific, the evidence gathered as part of this project suggests the importance of closed recruitment methods. The reason for this is simple—innovative knowledge is transferred this way. Indeed, as the demand for highly skilled and specialized staff increases, the use of networks becomes more important because the size of the potential pool of workers is very small. In this context, knowing who is innovative and who might be looking for a new opportunity requires insider status.

Terry: We are just about to employ someone here who is from Australia. He has relocated here with his wife and child and he has done [this work]. . . . [W]e proactively recruited him. . . . [W]e talked to some big companies who [do this kind of work]. We talked to contacts and said to them, "this is what we have got," do you know anyone who might be interested, who might want a change, even if it's only for three years? And this is how we learned about him. . . . Again word of mouth. . . . [I]t's interesting because the standard formal process in what I call these thin industries, don't really work. There's not a lot of people to choose from in New Zealand—even in Australia. The standard recruitment processes, we find, don't really work. A lot of it is to do with who you know in the industry.

Jo: We advertise positions, but generally we've got someone in mind. . . . It is a small country, and generally the cream rises to the top. You can usually find out who is good by asking round. We have a formal process, but invariably we have someone in mind.

Another reason networks were important in the recruitment process was that not all of the qualities needed to be successful in innovative firms were measured by educational qualifications, including those at the doctoral level. Thus, although potential recruits might complete high-level postgraduate training in relevant areas, merely holding a qualification was not sufficient to signal to managers that they necessarily had the skills required to be innovative. Managers reported that innovation depended on workers holding both technical skills and other dispositions and attitudes, including, for example, an ability to make connections between science and business and the ability to be a team player. Other forms of evidence were required to assess whether potential recruits had these qualities, and this was sourced through social networks.

> Tony: We have values and a culture, and it's important that [the new recruits] fit these. . . . [W]e ask them who they worked with, what compounds they've worked with. This might sound strange, but internationally, biology is a small field. Most people know someone who knows someone. . . . We tend to know who's in Australia and the United Kingdom. Europe, we're not so good at. And we know people on the west coast of the United States, because we have people in the field in these locations. . . . The thought leaders in . . . [our field], we know very well.

Although recruiting workers with specialist skills was ongoing in every phase, beyond the start-up and development phases, new human resource needs were identified. For example, in contrast to the needs present at earlier stages of development, during the production phase, biotechnology firms reported an increased need for technicians and a decreased need for scientists. At this stage, the process changed, and there is less emphasis on recruitment through networks and more emphasis on advertising and recruitment.

> Maddy: Our basic science has become less important as we have become more commercial. That's a basic model. We employed over 40 people here once. Of these, I'd say about 30 were scientists. Now we have something over 20, and 10 are scientists because the basic research changes into the "drugability" and end-use phase. . . . The science component of our business has become smaller and smaller—as we focus upon clinical trials, regulation drugability components have increased in significance. . . . This has changed the skill mix we require. . . . For example, we are just about to employ a document manager. Well you might say this sounds like a terribly bureaucratic position for an innovative biotech firm to offer. Well the regulators have such stringent requirements about record keeping that you have to have someone in that role.

Beyond the start-up stage, international networks continue to be of importance for other reasons. One was that there was a lack of local capacity. It was not that the knowledge was tacit and difficult to transmit in social ways; rather, it was simply that very few, if any, biotechnology firms had the experience needed to advance their products at subsequent stages of development. This weakness was compounded by a lack of training in the area in New Zealand. Thus, firms were reliant on workers from overseas who had the skills and experience needed.

Interviewer: Are these people likely to come from New Zealand universities?

Tom: Well, the proof is in the pudding. In terms of the scientists in our [research wing], the answer is yes and no. I'd say it's 50/50. In terms of the people here [the commercial side of the business], particularly . . . those who do the middle-phase work [drugability] and the end phase work [regulatory work] come from overseas mainly. [This is] because it is a young industry in New Zealand, and they have been doing it for 20 years in the United States—we've been doing it for 5.

Links with Tertiary Providers

So far, the data suggest that the most important linkage between biotechnology firms and providers of tertiary education rests in the discovery of new products that have commercial potential. Thus, linkages can be found at the start-up phase, where scientists working in universities and other creators of innovative knowledge form partnerships with others to commercialize ideas. Beyond this point, firms need to gain access to knowledge and skills that would advance their products. Advancing these products depends on gaining access to the latest ideas and most highly skilled workers, meaning recruitment needs to be global in focus. This wide focus was necessitated by a lack of local experience and a lack of capacity in New Zealand's tertiary education providers.

However, in contrast to the arguments outlined above concerning the flow of tacit knowledge and the development of institutional reputation, it does not appear to matter where the training was gained. So long as the employers had reliable evidence about the potential of new recruits concerning their training and their social skills, they were considered suitable employees. For example, one manager reported reviewing the Web site of the institution from which a potential recruit had graduated to assess the quality of his training.

It is important to note that not all universities have the capacity to train researchers in the skills needed. However, the managers' accounts did not evidence any reputational capital among those who had this capacity. So long as the potential recruits could demonstrate that they had the technical competency needed to be effective, could demonstrate they had the ability to see the commercial potential of their science, and had a favorable personal reputation, they were employable. In respect to this, although holding a doctorate in a relevant area (which could include chemistry, biology, physics, or cognate disciplines) was a prerequisite for entry into research positions, a doctorate was considered a mere introduction to the field, in that it demonstrated that applicants understood the rules of the process of scientific discovery. In other words, it was seen to give them a vocabulary, or the language needed to converse in the field. In general, there was nothing special about the qualifications, or the specific university from which potential recruits had gained them that influenced the managers' recruitment decisions. So long as applicants gained their degrees from research institutions, and they were in a relevant area, they had passed the first hurdle.

It is important to stress that concerns were expressed about the ability of such qualifications to signal the qualities and skills the managers sought. For example, managers reported that it was impossible to tell how much responsibility and initiative a candidate had taken when producing their own doctorate. Thus, it remained unclear whether or not potential recruits were actually innovative, even though they may have come from an institution that had a good record in research.

> Tom: Some people can get qualifications, but they are jolly hopeless. I don't think universities sort people out. I mean, some big players run factories producing PhDs. They turn out people who have the skills in some tiny areas, but they are completely unable to learn in some other area.

By publishing in top journals, potential recruits could increase their employability. However, researchers did not gain insider status in the field by simply attending particular institutions. They needed to move beyond the institution where they studied and build their own reputational capital. Recruits could build their reputation by participating in projects and by participating in the knowledge transfer ecologies that characterize the field. Some reported that individual mentors could be of significance in competitive spaces and that emerging researchers could build their own network capital by tapping into the networks of their mentors. Through a series of complex relationships, which involved participating in networks, the transference of skill and knowledge is best seen as an ecology. In other words, the transference of skill and knowledge is neither linear nor a one-way process.

Although remaining competitive in the field requires that managers maintain their global recruitment strategies, local developments were seen to hold out the promise of changes to the rules of advancement. For example, participants noted the growth in scientific capacity that had been encouraged by significant government expenditure on new equipment, such as the recently announced investment of NZ$10 million in the Institute for Innovation in Biotechnology. The Institute for Innovation in Biotechnology is also significant because it provides opportunities for colocation so that any barriers that currently exist between the university of Auckland and firms are further eroded. Thus, the university is also trying to assist the field by creating boundary spanners for those people who understand the commercial imperatives under which firms operate but who also understand biological process through their postgraduate degrees in bioscience enterprise. An important feature of the program is the development of network capital so that graduates have access to employers as part of their training. The extent to which the program succeeds in creating a portal into the field for graduates of the university will need to be assessed once students complete their training. Nevertheless, the program does demonstrate an appreciation of the importance of network capital by the university. This book returns to this idea in the concluding chapter.

Conclusion

Zucker and Darby (1996) argue on the basis of their examination of the biotechnology sector in the United States that for at least the first 10–15 years of the sector's development, knowledge of the innovations was held by a small group of discoverers, their coworkers, and others who gained knowledge while working directly with these workers. Ultimately, this knowledge spread out into other universities and became part of the routine science taught in most major research universities. Initially, the returns to scientists who understood the technology were high. However, as the knowledge spread, its value declined. Over time, the routine science only had value when incorporated with new research techniques in the most promising areas. Because these scientists tended to be reluctant to leave their then-current university positions and research teams, their own geographic location heavily influenced where development in biotechnology took place.

The findings of Zucker and Darby (1996) hold relevance to our understanding of the evolution of biotechnology in New Zealand. It is clear from the data that the development of new biotechnology products occurs because of a star scientist. It is also clear, however, that further developing the innovation relies on gaining international sources of knowledge. The data show that

New Zealand does not usually have the specialist skills required to undertake certain work in the sector. Developments in the field are occurring globally, and it is through making use of these developments that the firms can progress. To be sure, the firms do make use of local sources of knowledge, but these sources are best understood as additive rather than primary.

The literature and the evidence gathered in the biotechnology case draw strong links between innovation and partnering. The data show that partnering is needed both to gain access to new forms of technology developed in universities (particularly at the start-up and the development phases) and to gain access to forms of financial capital (at all phases). In terms of gaining access to capital, an ability to span boundaries was needed. The ability to present scientific ideas in a credible and accessible manner to investors was particularly important.

At various phases of production—from start-up through to full production–networks also played a role in the recruitment of staff. At the start-up phase, the interviewees reported that finding scientists who had the ability to innovate required being well embedded in the field. The highly specialized nature of the work meant that there was not a wide pool of suitable candidates, and recruiters often relied on word of mouth to identify suitable recruits. Even where positions were formally advertised, recruiters reported they usually had candidates lined up for positions. Although holding high-level qualifications (usually doctoral level) was a basic requirement, interviewees reported that the potential to be innovative was poorly measured by qualifications. Thus, potential recruits needed to hold high-level qualifications relevant to the area in which they would eventually work and to have good social networks.

Research universities and other institutions that are strong in biotechnology in New Zealand, such as the University of Auckland, may provide both the initial innovative idea and the individuals who initially champion these ideas. However, beyond this stage, the evidence gathered so far suggests that there is little that is unique provided by this university, in terms of skilled labor, that can not be sourced elsewhere. Indeed, recruitment tends to take on a global—rather than local—character, with interviewees reporting the quality of overseas graduates in some areas as being of superior quality. Thus, although the University of Auckland might be developing its capacity and may increase its reputational capital (Brown and Hesketh 2004), the evidence gathered so far in this research project suggests that the network effects are limited to the start-up phase, when the initial ideas are commercialized.

CHAPTER 7

Innovation and Networks in the Knowledge Economy

The Case of Knowledge Transfer Partnerships in England

The [Department for Trade and Industry]'s Innovation Review identified access to Partnerships and sources of new knowledge as two of the most important determinants of business innovation performance. Because innovation is a complex process, success relies on the coming together of a variety of players, such as suppliers, customers, other firms, universities, research and technology organisations and other intermediaries. Together, these players form part of the knowledge transfer system. (Department for Trade and Industry 2007a)

Introduction

As discussed throughout this book, the potential role universities play in supporting innovation has been recognized by governments keen to build knowledge economies. Indeed, as part of increasing their investment in knowledge creation and transfer, policy makers and politicians in the United Kingdom have focused their attention on increasing the links between universities and other providers of innovative knowledge, and industry. For example, in the United Kingdom, the recently implemented Science and Innovation Investment Framework for the period 2004–2014 is based on the notion that

[H]arnessing innovation in Britain is key to improving the country's future wealth creation prospects. . . . (Britain) must invest more strongly than in the past in its knowledge base, and translate this knowledge more effectively into business and public service innovation. Securing the growth and continued excellence of the . . . public science and research base will provide the platform for successful innovation by business and public services. (H.M. Treasury 2004, 10)

This chapter focuses on knowledge transfer partnerships that have been established in the United Kingdom to increase innovation. Central to the intervention is the use of associates, or individuals employed by the providers of innovative knowledge (or knowledge bases, as they are termed in the official documents), as conduits of innovative knowledge. The chapter aims to explore knowledge transfer partnerships to understand more clearly the way managers in firms, tertiary education institutions, and job seekers develop and use networks to advance their interests. Thus, as in the previous chapters, the purpose of this chapter is to assess the relationship between education and training and innovation and to uncover how the labor market for skill and innovative knowledge functions within a predefined field. In contrast to the previous two chapters, however, in this chapter the focus is on direct forms of intervention by the state in network formation, and the focus shifts to the United Kingdom, where New Labour has set about actively creating networks in some fields. The formation of knowledge transfer partnerships represents a clear example of this and thus provides an opportunity to explore how processes of knowledge transfer affect the competition for advancement through tertiary education. In turn, the findings are used to speak to broader network processes that may be occurring within the diverse range of university–business interactions identified by Hughes (2006). The case selected is the design knowledge transfer partnerships.[1] This knowledge transfer partnership has been established to help companies that would benefit from improved design solutions.

In laying out the case, the chapter proceeds through the following sections. Section one draws on Hughes (2006) to outline the nature of the interactions that occur between universities and industry in the United Kingdom. The next section describes in more detail the structure and role of knowledge transfer partnerships and situates them within New Labour's policy framework. This section also briefly reviews the little research that has been undertaken into knowledge transfer partnerships and their predecessor, the Company Teaching Scheme.[2] Section three outlines the method used to generate the results. The following sections draw on interview data from interviews with those involved in knowledge transfer partnerships in the selected field. The data are used to assess the central research questions posed earlier.

University—Industry Interactions in the United Kingdom

As Hughes (2006) argues, there is much evidence suggesting that the interactions and linkages between universities and industry are diverse. To understand the apparent diversity, he identifies four main kinds of relationship. First, there is a relationship between education and training: those who hold qualifications earn more than those who do not. However, the precise nature of the relationship remains unclear, particularly in fields where the qualifications produced in our institutions of tertiary education and training appear unrelated to the work done (see Chapter 5). Second, universities increase the stock of codified knowledge; for example, knowledge published in academic journals and the like that firms can access. Again, the extent to which universities interact with firms in this way varies across the sector and is influenced by educational reform. For example, as argued throughout this book, reform designed to concentrate research capacity in relatively few institutions will arguably also concentrate the production of tacit and codified knowledge in these institutions. Third, Hughes identifies a relationship between universities and industry that revolves around universities problem solving in relation to specific business needs. Here, activities such as consulting come to mind. Finally, he identifies relationships that develop as a result of universities' public space functions. These include a broad range of relationships, from informal social interactions, conferences, and specifically convened centers to promote, for instance, entrepreneurship, through to the exchange of personnel, including that which occurs through internships.

When considering these relationships, it is important to remember that the stress placed on these various functions by different universities varies across time and across space. One factor that affects the nature of the relationships between universities and firms is the nature of the systems of innovation in which they participate. Given that considerable variation occurs between different kinds of innovation systems, it is not surprising that the nature of the relationships between universities and businesses also varies considerably. Indeed, as Lester (2003) has shown, particular pathways to innovation exist for companies that compete through the use of leading-edge scientific and technological innovations. In such instances, the relationship between tertiary education organizations and employers is likely to be characterized by collaborative research and development clusters and through the role that providers of tertiary education play in producing graduates that have the technical and scientific skills firms require to innovate. For other companies, the pathway to innovation is derived through sourcing human capital with the right aesthetic qualities (including charisma). In such instances, positive relationships between firms and providers of tertiary education are more likely to arise if providers of tertiary education are able to recruit students who already hold the aesthetic skills needed to innovate. Of course, as is shown later in this chapter, it is not an either/or situation, and in many contexts, advanced technical and scientific knowledge and social and other skills are needed to promote innovation.

In a final example, where regions (and the companies within them) are attempting to modernize their focus, relationships with providers of tertiary education might be characterized by attempts to create bridging social capital. In such instances, firms might attempt to push out beyond their local horizons (by, e.g., bypassing their local universities) in order to tap into national, if not global, knowledge flows. In doing so, they might need to distance themselves from narrow or parochial relationships that might exist locally. Hughes (2006) stresses that the variety of interrelationships available allows a rich set of possible patterns of interaction and that there exists no one best or true way for universities and companies to interact.

A further contribution of Hughes' (2006) research is that it documents the sources of companies' knowledge that led to innovative activity. In his survey of firms in the United States and the United Kingdom, Hughes found businesses to be engaging with universities in a wide range of ways. In terms of frequency of engagement, informal contact with universities was the most common interaction that led to innovation. This was closely followed by graduate recruitment, use of publications, and attending conferences. At the other end of the scale, licensing and patenting were among the least frequently cited interactions that contribute to innovative activity. However, overall, when asked to describe more broadly their sources of knowledge that led to innovation, businesses ranked universities very low in frequency of use. Other sources of knowledge were seen to be more important. For example, respondents to Hughes' (2006) survey identified customers, suppliers, competitors, and internal knowledge within the organization as the dominant sources of knowledge for innovation. Finally, the results of the survey confirmed a point raised in the Lambert Review of business–university linkages (H.M. Treasury 2003), that the major challenge facing the effective exchange of knowledge between universities and industry was increasing business demand for research from all sources, including universities. Indeed, the Lambert Review made a strong plea for policies that promoted increased interaction between universities and firms.

The introduction of knowledge transfer partnerships represents an attempt to increase the level and the quality of university–business interactions. The next section discusses this intervention in greater detail.

Knowledge Transfer Partnerships

As described in previous chapters, in the United Kingdom, New Labour is encouraging providers of innovative knowledge to develop linkages and participate in partnerships with firms (Howells and Nedeva 2003). For example, in the proposed new skills academies, New Labour hopes to increase the links between secondary-level education schools and firms. The hope is that this will lift the skills of senior school students and increase the ability of firms to innovate through fostering innovation and spreading best practice. Similarly, linking centers of vocational excellence (which aim to increase the delivery of specialist work-based learning) with universities, training providers, and specialist schools will help construct networks in which employers participate in each sector (H.M. Government 2005). Other developments, which are designed to increase standards across the system by fostering innovation and spreading best practice throughout the economy through network formation, are apparent in both the priority partnerships of the Economic and Social Research Council and in the Higher Education Innovation Fund, which is providing £185m to help universities develop closer links with business and industry in the wider community.

The Department of Trade and Industry has also developed knowledge transfer networks, which consist of groups of knowledge transfer organizations. Similar to other network strategies, knowledge transfer partnerships are intended to strengthen the relationship between firms and knowledge creators in certain areas in relevant technologies. They are available throughout the United Kingdom and form part of the technology program. The technology program is one of the Department for Business, Enterprise and Regulatory Reform's business support solutions and is designed to stimulate innovation in the U.K. economy through higher levels of research and development and knowledge transfer.

The overall aim of knowledge transfer network support is to improve knowledge transfer into businesses. There are three main elements: managed networks, which aim to encourage the exchange of knowledge and information by way of outreach activities; information networks, which are designed to foster cross-sector and cross-border debate by way of signposting activities; and issues networks, which seek to draw industry players together to carry out problem-solving activities.

A related product in the Department of Trade and Industry's range of strategies to drive up innovative capacity is knowledge transfer partnerships. Similar to knowledge transfer networks, the idea is that knowledge transfer partnerships will drive up innovation by helping to develop pools of knowledge and by fostering nationally greater collaboration between firms and knowledge creators across the nation. Similar to other products in the Department of Trade and Industry's technology program,[3] knowledge transfer partnerships are offered throughout the United Kingdom. They are designed to stimulate innovation by speeding up processes of knowledge transfer and encouraging research and development designed to enhance the exchange of knowledge and information by way of outreach activities that foster cross-sector and cross-border debate and problem solving. There are currently over 1,000 knowledge transfer partnership projects underway in over 15 discrete areas. These areas include aerospace and defense, bioscience, chemistry, and food processing (for a full list of knowledge transfer partnerships, see H.M. Treasury 2003). The total budget for the partnerships is in excess of £20m per annum.

Similar to the other products in the Department of Trade and Industry's range, knowledge transfer partnerships are based on the premise that fundamental research is essential but that the knowledge generated must have ways of being converted into products that increase competitive advantage. To date, there has been much research that has the potential to contribute to achieving this goal. However, this potential has yet to be realized, as the knowledge has tended to remain trapped in the heads of researchers or the journals in which they publish their findings. Thus, as Hughes' (2006) research suggests, the underlying logic is that the potential of many research discoveries remains unrealized because researchers do not properly appreciate their commercial possibilities, they lack access to forms of capital needed to commercialize their ideas, or they lack the inclination to partner with other groups and organizations that might be able to advance the use of these ideas. In a similar vein, many businesses lack knowledge of discoveries made by researchers that can be used to increase their capacity to be innovative. Indeed, a major strength of arguments justifying the formation of knowledge transfer partnerships is that it explicitly recognizes that the complexity of innovation is such that it relies on a variety of players coming together to transfer knowledge and build relationships. In addition, it establishes that networking is critical to the process.

To receive government funding, companies must choose a partner from the U.K. knowledge base (such as a university or a college). Together, the host organisation and the knowledge base prepare a proposal for a project that will enhance knowledge transfer; this is then assessed by the Partnerships Approvals Group (which comprises the Department of Trade and Industry and industry representatives).

As noted, central to the scheme is the use of associates and their supervisors, who effectively provide linkages between the knowledge bases and participating companies. Associates are employed by the knowledge bases but work in participating companies to complete projects under the supervision of a senior researcher. They provide a key means of transmitting the knowledge discovered and the solutions to the technical, scientific, and social problems found in the knowledge base into firms. Thus, participating firms can access two sources of knowledge—that provided by the researcher and that provided by the associate.

The formation of partnerships has been aided by the establishment of knowledge transfer partnership offices in many knowledge bases. The purpose of these offices is to facilitate university–business links and to generate revenue for the universities in which they are based. Most knowledge transfer partnerships generate a small but important surplus for universities. There are other benefits for participating universities, such as an increased ability to fund research. The main expense for companies is the cost of employing the associate undertaking the placement (though there may be substantial hidden costs, such as new equipment that firms may need to purchase). For smaller companies (fewer than 250 employees), the grant covers about 60 percent of the labor costs (about £16,000–18,000 per annum). For larger companies, usually 40 percent of the cost is covered.

To date, there has been little empirical research into knowledge transfer partnerships, nor have they attracted much critical interest from researchers, though this is beginning to change as the intervention takes effect. In part, this reflects the newness of the scheme. Nevertheless, some information about the effectiveness of the scheme can be gleaned from official reports, which suggest that participation in knowledge transfer partnerships offers significant financial returns to participating firms. For example, during the 2005–2006 financial year, firms involved in partnerships reported an average one-off increase in profit before tax of £78,000. After the projects had been completed, participating companies reported an increase in annual profit before tax of £291,000 (Department of Trade and Industry 2007b). The Department of Trade and Industry also reports that for every £1m the government spent on the scheme, the average benefits to participating firms amounted to a £4.25m annual increase in profit before tax; £3.25m investment in plant and machinery, with 112 new jobs created; and 214 staff in the firm trained as a direct result of the knowledge transfer partnership. For the knowledge base, the benefits of participation include (on average) 3.6 new research projects and two research papers. For the associate, the Department of Trade and Industry reports that over 60 percent are offered and accepted a post in their host company on completion of their project. It is noteworthy that the associate may not come from the knowledge base that has formed the partnership with a firm in the first instance. Rather, all associate positions must be advertised and open procedures adhered to (i.e., advertisement and interview). Finally, 41 percent of associates are registered for a higher degree during the time they are involved in the knowledge transfer partnership, with 67 percent of these ultimately completing their studies (Department of Trade and Industry 2007a). On the basis of this statistic, other information, and the findings of the Lambert Review (H.M. Treasury 2003), the size of the funding pool available for new partnerships continues to be increased.

As noted, there is a dearth of academic research into the knowledge transfer partnerships. What little has been written tends to describe small case studies that identify the advantages to firms from participation (see, e.g., Wynn and Jones 2006; Wynn et al. 2008). In addition, there has been an official evaluation into the Company Teaching Scheme, from which knowledge transfer partnerships evolved (there has been little, if any, academic research specifically exploring the Company Teaching Scheme).

One of the testable hypotheses of this book is that the creation of innovative knowledge would be concentrated in elite institutions, such as those that score highly in the Research Assessment Exercise. The findings of the official evaluation do not lend support to this assertion, at least as it relates to the Company Teaching Scheme. The evaluation found that 60 percent of knowledge base partners had Research Assessment Exercise ratings of 3 or lower, indicating low research capability and nonelite status. A significant minority (12 percent) had ratings of 5 or 5* (Department for Business, Enterprise and Regulatory Reform 2002). Although other interpretations are possible, given that over 90 percent of all applications are successful, the high proportion of lower-ranked institutions participating in the scheme suggests that innovative knowledge is created by a diversity of institutions.[4]

In other findings, in terms of the firms involved, the Department for Business, Enterprise and Regulatory Reform (2002) evaluation found that over 90 percent of partnerships were with small to medium enterprises. A minority of firms (38 percent) of those surveyed reported that the technology/knowledge transferred into their firms by the partnerships was new to them. A further 45 percent of interviewees reported that although they had knowledge of the technology, the work the partnerships allowed represented a considerable advance for them. The principal motivations for the knowledge base partners to participate were first, a desire to see knowledge created in their institutions transferred into industry, and second, a desire to enhance their research and teaching. Finally, the role of the lead academic was identified in the evaluation as being critical to the success of the scheme.

Having outlined the nature and effect of the knowledge transfer partnerships, attention can now be focused on the method used to gather the data reported later in the chapter.

Method

A selection of senior academics and employers (referred to here as managers) who participated in a design knowledge transfer partnership was interviewed as part of this case study. The majority of informants were based in England. However, a small number were based in other parts of the world, where the universities they worked for provided outreach services. As is the case with the previous two chapters, the interviews were designed to map the human resource practices within a field, with a view to assessing the rules of advancement in operation in the field. The Department of Trade and Industry's knowledge transfer partnership Web site provided details on those participating in the scheme. After identifying potential recruits at random and seeking their support by e-mail, telephone interviews were undertaken.

A qualitative methodology was employed based on semistructured telephone interviews (ten with managers and eight with associates). The interviews were designed to gather a range of data including those pertaining to the role played by managers in identifying areas worthy of applying for a knowledge transfer partnership, as well as their motivations for participating in the scheme. In addition, informants were asked about the strategies they employed when selecting associates. Where necessary, the interviewees were asked additional questions via e-mail. As part of the ethical considerations that underpinned this study, all participants were told that the purpose of the research was to explore the functioning of the knowledge transfer partnership. Participants were guaranteed anonymity, informed that they could withdraw from the study at any time without reason, and told they could withdraw any data already provided. Pseudonyms are used to protect the identities of all participants and their organizations. As soon as practicable after interviewing, relevant data were transcribed. Relevant data were those that were considered central to the enterprises and individuals satisfying their human resource requirements (including those related to skill demand). Data analysis involved identifying the key themes and experiences of each interviewee. To improve the readability of the data, some text has been removed and some added. In this respect, an ellipsis (. . .) indicates text that has been removed, and square brackets ([text]) are used to indicate text that has been added. The abbreviation KB (knowledge base) is used to indicate the providers of the innovative knowledge. The knowledge base's most recent rating in the Research Assessment Exercise is also reported. An early draft of the case study was returned to the participants so that they could attest to its accuracy. Several informants suggested changes to the text. All of these changes were made. It is important to acknowledge that this study explores the practices of those participating in the design knowledge transfer partnership. Caution is required when generalizing these findings broadly to other knowledge transfer partnerships.

Knowledge Transfer Partnerships as Conduits for Innovative Knowledge

As shown throughout this book, previous research conducted in the United Kingdom and elsewhere has highlighted the importance of social networks to processes of knowledge transfer. The creation of knowledge transfer partnerships is designed to help create these networks by supporting the creation of a social infrastructure conducive to the transmission of innovative knowledge. The key themes that emerged from the data were sources of innovative knowledge and selection and recruitment of associates. These themes are discussed in turn.

Sources of Innovative Knowledge

It became clear early on in the interviews that the knowledge bases had both knowledge that would help drive improvements in production and the skills needed to find solutions to technical challenges faced by firms. Thus, the contribution made by the knowledge transfer partnerships to innovation was not just that they provided a way to transfer innovative knowledge that had been created in knowledge bases directly into firms but also that, by encouraging the formation of learning communities, they facilitated problem solving. To ensure that the knowledge transfer partnerships contributed to innovation in these ways, the managers interviewed as part of this study reported that they needed people who had both technical expertise and the other skills, such as the ability to solve problems. Bringing this expertise and these skills to bear on a design problem was perceived to speed up development of new ideas and to reduce inefficiencies through, for example, eradicating redundant information. The data show that redundant knowledge was also to be found in academic papers. This suggests that unless employees have the means to decode and understand the value of the knowledge contained in codified forms, such as that contained in research papers and the like, access to such knowledge is unlikely to help them improve their competitiveness. Rather, to use codified data, firms need insiders who can decode this knowledge and identify its value to firms.

Sandy: I think the beauty of the knowledge transfer partnership is that
(KB RAE 5)   it stops the companies doing years upon years of work doing
             stuff that is never going to work. It eliminates a pile of work,

because of the knowledge those involved in the partnership have. Normally . . . [the firms] want to develop something new, so there is never any guarantee of success. So really what you are saying is the person in the university has the background knowledge needed to know what is or is not going to work. You can then concentrate on stuff that is going to work. That's the real knowledge transfer I think. . . . You take the original ideas from the discussion and you say, "OK I can see why you want to do it, but you can't do it that way, you have to do it this way." So you come to a logical path, which is far faster than if they had been left to their own devices. That it is the real knowledge transfer—getting rid of the nonsense that is hard to filter out of the literature. Because people publish a lot of stuff in the literature, and unless you've got a guru to tell you who is good and who is bad and what's rubbish and what's not, it is impossible for someone new to the field to decide. You need someone who has been around for 10 years or so in certain areas.

Interviewer: So even the literature is not always right?

Sandy: Far from it. Someone who's been in the field for 10 years and is eminent would have a very critical understanding of the literature and a level of believability that you could assign to any given paper or idea in the paper. That really is the knowledge transfer—that's the thing [that makes the knowledge transfer partnerships work]—it's the background knowledge of the academic partner.

As noted, it was not always the case that the knowledge had already been created in the knowledge bases and was simply waiting to be transmitted into the firms. Rather, respondents reported that they participated in learning communities. In this respect, the source of the problems associates were employed to tackle was not always obvious and required detailed exploration and investigation. Thus, in some cases, innovation was an interactive process involving the associates, their supervisors from the knowledge bases, and actors from the host firms.

Pam: We were looking for ways to help Jones improve the soil sustain-
(KB RAE 5)   ability and improve their production. We started by looking at the actual soil. We looked at the actual mechanisms at work—how they tilled the soil. We then looked at the machines that go on the field. We then looked at the actual machine that does the work with the soil. Because that was what we found was making the most impact on the soil. We then contacted the manufacturer of the machine to see if there was anyway we could redesign

it, so that we could reduce its impact. From there we managed to get an associate to work with the manufacturer on improving the machinery. . . . What was interesting about the project was the way different parties fed information back to the manufacturer of the machine. I had a number of research projects with Jones. Once I had the results of these, which identified ways to improve production, these were fed back into the manufacturer of the machine. There'd be meetings with them, where we'd discuss what had happened in the field. Grower groups would also attend meetings with Jones with a view to driving improvements to the machine.

An additional aspect of processes related to innovation was the way reputation underpinned the formation of the knowledge transfer partnerships in the first instance. In a number of cases, particularly where highly specialized scientific and technical knowledge was sought, the knowledge bases reported that they had standing in their fields, and it was this reputation that helped to bring the knowledge transfer partnerships into reality. Though there are methodological reasons to remain cautious of such self-affirmation,[5] in some instances, reputation was based solely on intellectual resources. In others, both knowledge and the other resources the firms could access via a knowledge transfer partnership (e.g., plant and equipment) were the sources of reputation. Whatever the case, when asked what motivated the firms to participate, both the knowledge bases and the managers reported that the reputation of the knowledge bases was an important motivating factor.

Interviewer: Why did the firm come to you?
Henry: We are a very big research group with a good reputation in the
(KB RAE 5)   area. We've developed certain kind of knowledge, which is very
              advanced. The knowledge is very specialized and is only available
              from this university. The company wanted this knowledge to
              increase its competitiveness. It needs the knowledge to compete.
              The international market in the area the company works in is
              very competitive. It needed access to the knowledge to boost its
              competitiveness globally and enhance its reputation.
Interviewer: How did they know you had the knowledge?
Henry: I am part of a big research group in the area, and we have a
              strong reputation in the area. We've also entered into collabora-
              tions with them in the past.

Interviewer: Could the firm have got the knowledge it needed to advance its
              products from elsewhere?

Joseph: Well, you're asking me to get a bit cocky here, but there's not
(KB RAE 5) a lot of people who have the expertise. First of all, in the pro-
            teins community, my technical knowledge is very good. I've a
            quite a few of the leading papers in the field. So that's an obvious
            thing. Second, on the emulsifiers side of things—this is where
            they wanted to go, and my university has a massive reputation
            in the area. There are other places that do it, but I can't think
            of anywhere that has an interest in the two areas. Some people
            could work in both areas, but not very many. So it would be hard
            to go elsewhere. Even on a world scale, even if there had been a
            choice of anywhere of the world [which there wasn't because of
            the regulations governing the scheme], my university would have
            been at the top of the choices.
Interviewer: How did your university get the reputation?
Joseph: That was not my doing. It goes back 70 or 80 years. In proteins
            it goes back 70 years and goes back to the work of Professor
            Cavalle in the 1930s. . . . [T]he basic understanding on the pro-
            tein side of things goes back years. We build up a whole suite of
            equipment for protein manipulation that is second to none in
            the academic community in the United Kingdom.
Interviewer: So did the firm come to you because of your reputation?
Joseph: Well, perhaps not for the emulsifier side, but certainly they
            would have come to this university because of our reputation on
            the protein side, but it was a happy coincidence that we are also
            at the top of the field in emulsifiers.

These data are interesting because they show how reputation is related to
the levels of investment in plant and equipment and to the level of expertise
researcher had in their fields. In this respect, it is possible to argue that the
concentration of research and development capacity in relatively few elite
institutions will limit the ability of nonelite institutions to develop reputations in fields where the costs of establishing a reputation are high. The extent
to which this concentration of capacity helps create pathways for some into
elite positions in the labor market will be considered shortly. For the present, it is important to point out that although the reputation of the knowledge bases provided the impetus for some of the firms to seek out knowledge
transfer partnerships, it was possible for knowledge bases to establish knowledge transfer partnerships in other ways. This is particularly the case where
the knowledge transferred was not positional in nature (i.e., it could be
gained form different institutions of tertiary education). For example, in
some cases partnerships resulted from chance encounters between firms and
knowledge bases and as a consequence of knowledge transfer offices in the
knowledge bases seeking out partnerships with firms.

Sally: I think [the firm] came to one of the events put on by the
(KB RAE 3)   university to promote knowledge transfer partnerships. . . . I was
                not directly involved at this stage. I was only bought into the
                project once they were aware of this opportunity. Quite often
                this is the issue—you are making small businesses aware of the
                opportunity offered by the knowledge transfer partnership.

Interviewer: So this was a seminar that was run by the knowledge transfer
                partnership office at your university.

Sally: Yes. And quite often it is word of mouth so they hear about the
                opportunity through other people who get in touch with us.

Interviewer: Presumably your Knowledge Transfer Partnership Office is try-
                ing to advocate with firms for your university . . . to convince the
                firms that your university has the skills that they need to solve
                their problems.

Sally: Yes, but in reality, if the firms don't feel we have that—we're just
                asking for trouble if we try and engage with them but we don't
                have the resources to support them. If they have an enquiry that
                comes in, and they ask the question, "which discipline or which
                part of the university is most appropriate to this?" Sometimes
                I will get involved in going and looking and seeing to actually
                say, "well, I don't think this is operations, it's more marketing,
                or it's more computing." And you encourage them to go and
                find someone in those disciplines in the university who might be
                interested in a project. . . . In specialist areas, we might not be in
                a position within this university to support that project.

The data show that in areas on the economy where innovation is driven by
highly specialized scientific and technological knowledge, particular knowledge bases had reputations in the field that helped to bring about the knowledge transfer partnership. One source of this reputation was the fact that
the knowledge and other resources were only available from these knowledge bases. However, as noted earlier, not all knowledge transfer partnerships involved knowledge bases that had unique or positional knowledge. In
this respect, although all the interviewees reported that they benefited from
the knowledge transfer partnership, in some areas of innovation, firms could
have gained access to the skills and knowledge needed from a number of different knowledge bases. As the following data attest, in part, this reflects the
nonspecialist nature of the knowledge and skills they needed to access to be
innovative. In such cases, the proximity of the firms to the knowledge bases
and the activities of the knowledge transfer offices in setting up partnerships
were more important than any positional knowledge the knowledge bases
might have access to.

Interviewer: So why did Smiths Foodstuffs come to you?

Mike: Well in this case, it's because we are adjacent to them. They could
(KB RAE 3) have gone to Southern University. I suppose there's a natural
geographical link—you would generally go to somewhere that
is reasonably close. I suppose we have knowledge transfer partnerships where we travel maybe 30 miles, but this would be the
most—20 miles or less is more common.

Interviewer: So the transfers that your university offers tend to be local?

Mike: Yes, unless there is a special need to go further and you can identify us as provider to fill a special need.

Interviewer: You say they could have gone to Southern?

Mike: Yes—nearly all the universities run knowledge transfer programs.
. . . Quite often people might come because they have heard that
we have done work in the area before and those involved have
been quite pleased with it and therefore they come to us. No
doubt this has an influence even if they have not come by that
route because they are trying to assess which university they go
to, to have this knowledge transfer. Quite often I think the rationale behind that is not so much around the particular knowledge
base, because quite often they don't know what they want in the
first place, so they come to us for that advice.

Finally, the interviews with the data suggest that one barrier to cooperation
between firms and knowledge bases, particularly in areas where the knowledge
transfer partnership has been established to solve advance scientific and technical problems, is concern about protecting the intellectual property rights.
It was not simply that the knowledge bases had contained all of the positional knowledge. Rather, in some cases, firms exposed themselves to some
risk because to gain the most from the knowledge transfer partnership, they
had to reveal some of their own positional knowledge to the knowledge bases.
In some cases, knowledge transfer partnerships were only possible because
of the agreement that the knowledge base would not reveal the positional
knowledge. When asked why they had formed a knowledge transfer partnership with Thompson's, rather than with two other providers of innovative
knowledge that were also being considered, a manager replied:

Jeremy: Thompson's university gave me the best agreement over Inte-
(electronics llectual Property. This is a big problem. When you're a company
firm) you answer to your shareholders. You basically have to hold the
intellectual property. Your shareholders do not want you share
your intellectual property with the university. The other problem

with sharing your intellectual property with the universities is that they don't have sufficient funds to achieve international patent protections. So they are not even able to play the game even if they wanted to. . . . I need to have the right to protect any intellectual property that arises from our research and development activities. It doesn't matter if the intellectual property is created in a university, I still want the right to own the intellectual property. Thompson's was willing to give me that right, but . . . universities in the United Kingdom are increasingly unwilling to. Dunmore, Xavier, and other universities wouldn't. The other honest answer is that in the field that Sam [from Thompson's] is in, he's just the best. He's the strongest scientist in the field.

An additional idea this book seeks to evaluate is the extent to which the human resource practices employed by managers might help explain how institutions of tertiary education help reproduce social divisions. Brown and Hesketh (2004) argue that attending institutions richly endowed in reputational capital is important to individual mobility. The evidence thus far in this chapter suggests that in areas of the economy that demand access to positional knowledge, reputation in the field helps guide some firms' choices about with whom to form knowledge transfer partnerships. However, in areas where the knowledge needed to promote innovation is relatively widely available, there is a local dimension, with firms forming partnerships with local universities. The next section explores the human resource practices employed by the knowledge transfer partnerships.

Recruitment of the Associates

As noted above, the knowledge transfer partnerships are required by the Department of Trade and Industry to use open recruitment methods. In addition, because the associates were employed by the knowledge bases (typically universities), they had to adhere to the universities' human resource practices. Without exception, then, the knowledge transfer partnerships used open advertisement and interview to attract individuals to the positions and to select successful candidates. However, the way these practices were used varied across the knowledge transfer partnerships. The variations reflected a number of factors including the nature of the knowledge and skills the associates needed to successfully complete the partnership and the availability of individuals who had the knowledge and the skills needed in the labor market.

In terms of the first factor—the knowledge needed to successfully complete the project—the projects varied in the extent to which high levels of technical expertise and skill were needed. In some partnerships, the interviewees reported that the skills needed by associates were both technical and scientific, and they were soft. In terms of scientific and technical skill, it was not always essential that the associates held top-level or even high-level qualifications. This meant there were many applicants who met the basic requirements in terms of formal qualifications. For example, some reported that they had employed people who had upper-second-class honors in a relevant area. However, finding individuals who had the technical and scientific skill and other qualities needed to be effective in the knowledge transfer partnerships, such as the ability to use their technical and scientific skills to solve practical problems, was not so easy. In this respect, most of the interviewees reported that the associates needed to have distinct skill sets, including an ability to span boundaries between academic learning and practice. As the following data suggest, this created difficulties, as academic qualifications were not good proxies for the soft skills needed to be successful in knowledge transfer partnerships.

Interviewer:	How did you recruit the associate?
Tom: (KB RAE 3)	Well, the positions are advertised, and we decide as a company and as a university where they are advertised. Sometimes nationally, sometimes locally. Sometimes we get a very good response if we go to our alumni list. In recent cases, over 50 percent of associates are from our alumni who are looking for another job. We send them a letter saying "would you be interested?" So it's a case of creating the interest rather than actually responding to it. . . . [B]ut other people will actively be looking for associate positions, and therefore they will go to the . . . knowledge transfer partnership Web site and see what's there.
Interviewer:	In terms of the knowledge transfer partnership that you have currently with Sampson, how did you recruit the associate?
Tom:	He was one of our alumni.
Interviewer:	Did you know of him?
Tom:	I didn't, but my colleague who is from the computing department knew of him. But he did not specifically target him—he was just on the alumni list. We targeted people who had undertaken certain courses who had certain degree classifications.
Interviewer:	What level of qualification did you require?
Tom:	Well the usual minimum is a 2:1, but on occasions we have found that the best candidate may have a 2:2. As long as we have diligently selected them, we're allowed to accept them.
Interviewer:	So what other skills do you need?
Tom:	One of the big things is that they need to be self motivated and focused on the project. They also need to be fairly dynamic at

linking with people, because the project is about changing how people think and the way people work. It quite often involves working with people at all levels of the organization, from those on shop floor through to the managing director. We would like someone who is well equipped to do this. You don't want someone who is going to keep their nose in their computer. It is implementing the change rather than being merely a techie, stuck behind a computer.

Interviewer: Do you get that kind of information from their degree code?

Tom: No. . . . we try and make an assessment based on their previous experience to see if they are likely to have that personal motivation and personal management almost, and how they are likely to work with people in the organisation. Quite often that's a tricky area, and you are encouraging the associate and coaching them. You are encouraging the company to become involved. It's a learning process that does not come through degree work.

In other knowledge transfer partnerships, substantial postgraduate training in a specialized area was required (in addition to soft skills) to be effective in transferring the required knowledge. This meant that the size of the potential pool of applicants was relatively small, and applicants were likely to be known to those involved in the knowledge transfer partnerships.

Interviewer: Can you tell me about the process you went through to recruit the associate?

Terry: The university employs the associate. So we abided by all the
(KB RAE 5) university processes. The position was advertised, we short-listed, and then we interviewed. That's the process. We actually knew most of the applicants. Not personally, but by their work. This helped because we were able to judge whether or not they were representing themselves truthfully.

Interviewer: What signals of competency did you use?

Terry: We looked at the work they had done. . . . In this area, we needed someone with high levels of skill, and it would not be suitable for someone who has only relatively low levels of qualification. For this reason, we recruited a person who has a PhD. The skills needed were that advanced. . . . So they had to have an advanced qualification. We also looked at their publications and the kind of projects they had been involved in. The field is highly specialized, so we knew of the person's background and felt he could do the work. It is actually easy to judge if they could do the work.

In other cases, the managers reported recruiting those from within their research groups within their universities, indicating the use of network recruitment processes. The reasons for using network recruitment methods were twofold. First, the skills needed to be effective in the knowledge transfer partnerships were highly specialized. Only those who had been trained in the knowledge base or in similar programs (of which there were typically very few, if any) were seen to hold the skills and knowledge needed to be effective in the role. The second reason for recruiting insiders rested in the quality and trustworthiness of the information managers had about the candidates' potential. Managers reported that their detailed knowledge of those they had already worked with offered them a way of increasing the project's chances of success. This information was seen as being of far better quality than that available through advertisement and interview.

Interviewer:	Tell me about how you recruited them. For example, how did you choose the associate?
Joseph: (KB RAE 5)	[I]t was straightforward. I've had a large research group going for a number of years, and I chose two individuals who were well known to me who I've worked with, or whose studies I've supervised. So, I have two associates [on different projects at present]—one is a scientist who I worked with for his doctorate. And I have another who was supervised by a colleague, but who I also helped with his doctorate. I knew them both personally. I knew their capabilities and what they could achieve.
Interviewer:	You had to advertise the positions, did you not?
Joseph:	I think we did. It was along time ago, but I think we had to. These were the best candidates by a distance.
Interviewer:	Did you know the other people that were applying?
Joseph:	A couple of names were semifamiliar, but there was no one that was competitive with those that I eventually chose.
Interviewer:	How did those who were appointed learn about the positions?
Joseph:	I told them of the vacancies—personal knowledge far outweighs any knowledge you might get through an interview. I don't know how you can judge somebody's potential based on a CV and an interview. . . . You can eliminate complete no-no's, but to get the best candidate [through advertisement and interview] is very difficult. . . . I think recruiting someone you know works pretty well. You can overlook someone, and I suppose you could say it's unfair. However, at the end of the day, our job is make sure we get someone who is not going to be disastrous for the program and who can perform reasonably well. If you are confident that you have a person like that, then it's very hard to take a risk on someone else. You can ring up and get a recommendation, which I have done, just to make sure that my preferred person is the right choice. I've often found that it is better to go with who you know.

Interviewer: Is there anything particular about the knowledge?
Joseph: Oh yes, the knowledge and skill base has to match very well.
               Both candidates had the required knowledge of proteins and
               emulsifiers and the knowledge of the techniques. It's not just
               that I appointed people who I liked. That would be plain silly. It
               is critical that you choose good people . . . both are working out
               very well, and all are very happy with their performance.

Finally, it is useful to make two additional points. First, building reputations
through the placement of associates into firms led researchers to look for
recruits beyond their own universities. The researchers and the firms interviewed all reported that they wanted to select the best possible applicants as
associates. In some instances, building their own reputations in the field with
firms led researchers to pass over their own students and tap into national
talent when recruiting associates. Second, in some areas, providers of tertiary
education had strong reputations for expertise in particular areas but did not
offer training in these areas. In such contexts, the managers also needed tap
into national sources of labor.

Conclusions

The purpose of this chapter has been to use the case of knowledge transfer partnerships in the broad field of design to explore how managers make
and use networks to advance their interests. On the basis of the interviews
conducted, the following conclusions can be drawn: The data support social
network theory, but only in instances where the knowledge transferred is
highly specialized. For example, the data show that where firms seek knowledge that is not widely available, insider status was gained by those who had
been taught in the knowledge bases that had formed knowledge transfer
partnership sites. However, it was not that case that all knowledge transfer
partnerships involved the transfer of positional knowledge. There were many
examples where managers reported they only required graduates who had
sound training, as measured by their degree performance. There was little
evidence that the knowledge gained by these graduates in the studies was
unique to the university from which it was obtained. In other words, it was
not positional in nature, and graduates could gain the required training from
any number of providers of tertiary education. To be sure, the quality and the
depth of technical and scientific knowledge were important, but attendance
at a particular institution did not necessarily signal this.

Irrespective of the qualifications needed to gain a position in a knowledge transfer partnership, to win a position, all managers reported that the

candidates needed other skills and qualities. Many of these were not signaled by educational qualifications. All managers reported a desire for skills that were difficult to measure, such as entrepreneurship and self-motivation. For example, managers reported a need for associates to be at ease "on the shop floor" and in the lab. Gaining information about whether or not potential recruits held such qualities was difficult. In some instances, social networks provided a conduit through which this information flowed. In other instances, to make decisions, the managers relied on open recruitment methods and looked for other signals of capacity, such as the candidates' previous work experience.

These findings allow us to advance some arguments about the relationship between tertiary education and the competition for advancement. On the basis of the data, it seems that as the knowledge that is being transferred through knowledge transfer partnerships increases in complexity, and where higher-level qualifications are required (e.g., at the doctoral level), it becomes more likely that managers will use network recruitment methods. In other words, where capacity is thin, networks play an increased role in facilitating labor market relationships. In other contexts, networks may play a role; however, the fact that the scientific and technical knowledge needed to be effective is not positional in nature means that graduates from any number of tertiary education institutions can apply. The quality of the applicant's knowledge in the field and their social and other skills are the critical factors in determining employment.

From this point, two further points can be made. First the importance placed by managers on the technical and scientific skills and knowledge of the associates suggests that reputations can be enhanced through teaching quality. In turn, this may be enhanced by selecting into tertiary programs those students with the greatest propensity to learn. Second, reputations can be enhanced either by recruiting those who are best able to span boundaries between theory and practice or by teaching such skills to students.

CHAPTER 8

Summary and Conclusion

The central purpose of this book has been to assess the relationships between innovation, social networks, and the competition for advancement through tertiary education. Its aim was to contribute to our understanding by exploring some of the factors influencing network formation and by describing how changes in the provision of tertiary education help shape new rules of advancement. In addition to contributing to our understanding in these ways, this book has also attempted to provide a framework by which to understand more clearly processes of social reproduction occurring in tertiary education.

This chapter further assesses the key arguments advanced within this book by evaluating what light the empirical aspects shed on these and by considering implications for policy makers and researchers that arise from the research.

The Central Case

The central argument that has been advanced is that knowing what and knowing who are often perceived to be in conflict with each other. However, on the basis of the research presented in this and in other works, it can be seen that knowing what and knowing who are closely related to each other. Indeed, in some fields, changes in production are driving up the importance of knowing who as a means of gaining access to knowing what. For example, it is widely acknowledged that in knowledge economies, or those economies that derive competitiveness from innovative technological and scientific knowledge, ecologies of innovation are needed (Nahapiet and Ghoshal 2000). In such ecologies, there exists a complex interaction between knowing who and knowing what. In addition, this book argues that in many contexts, innovative knowledge is tacit in nature and, therefore, is best transmitted through social methods, including through social networks (or knowing who) and through labor mobility.

The relationships that exist among innovation, knowledge, and competitive advantage has led to emphasis in policy being placed on creating linkages between creators of innovative knowledge (particularly universities) and other actors who participate in knowledge ecologies. These actors include financiers and regulators, as well as individuals and the firms in which they work.

This book shows that policy makers in New Zealand and the United Kingdom have come to recognize the relationship between knowledge ecologies and competitive advantage and have set about designing policies that promote network formation. This is particularly the case in the United Kingdom, where significant investment in a range of knowledge transfer strategies has been made by New Labour and, as one respondent interviewed as part of this book put it, in the United Kingdom, knowledge transfer has become "big business." For their part, some universities have attempted to capitalize on opportunities to benefit from the reforms by establishing knowledge transfer offices.

Recent research has improved our understanding of the importance of social networks; nevertheless, there exists debate about whether or not government policy should be directed toward creating open or closed ecologies. In general, those interested in increasing equality of opportunity are committed to building open networks in which all individuals can participate. For example, Szreter (1998), who comes close to this position, argues for policies that encourage citizens to participate in the widest possible set of weak ties because they offer the greatest potential to challenge forms of social closure that have, to date, excluded nonelite groups from the benefit of economic growth. However, as shown earlier in this book, there are reasons to be less than sanguine about the value of open networks. Indeed, although this book accepts arguments made by some social capital theorists that sharing knowledge widely lifts the capacity to innovate in some contexts, it can be seen in many other cases that maintaining the contribution made by knowledge to innovation depends on knowledge not being widely shared, particularly with competitors. In other words, in many contexts, innovative knowledge has a positional aspect. Similar to many other works, then, this book has theorized that accessing positional knowledge provides a potent source of competitive advantage. One way to gain access to such knowledge is through forming linkages between creators of innovative knowledge (e.g., universities) and firms, and also in participating in broader knowledge ecologies.

Although policies developed by the state to assist in network formation makes economic sense in some contexts, this book has argued that social implications that arise from this intervention remain poorly understood. Because of this weakness, this book has argued that a better understanding of relationships between knowledge creation and knowledge transfer will improve our knowledge of both how competition for individual advancement through tertiary education is organized and how network processes contribute to new forms of social inclusion and exclusion.

One hypothesis advanced in this book is that those who attend institutions in which innovative knowledge is created are best placed to gain access to knowledge needed to win in the race for advantage, via education. Institutions that have the largest research and teaching budgets (per student) are best able to produce graduates who have the knowledge employers demand. It is noteworthy in this context that in 2001–2002, a massive discrepancy existed in the United Kingdom between the most prosperous mainstream university (Imperial College) and the poorest (Anglia Polytechnic University), with the former receiving over nine times the income of the latter. According to Brown (2003), this division in wealth helped to explain why Imperial College ranked much higher than Anglia Polytechnic University in the Times' League Table 2001.[1] In the table, Imperial College ranked third, whereas Anglia Polytechnic University ranked 97th.

By advancing the argument that it is knowledge created in research institutions that helps explain the formation of reputation, this book has attempted to provide a different way of looking at the competition for advancement through education than is presented elsewhere. Other accounts have argued it is cultural capital (expressed in the notion of personal capital) held by segments of the middle classes, rather than technical and scientific knowledge gained from tertiary study, that explains the formation of reputation. Brown and Hesketh (2004), for example, argued that personal capital, in forms such as charisma, are now more closely bound up with employability. An aspect to their argument is that massive growth in tertiary education participation has made it increasingly difficult for employers to have first-hand knowledge of universities or the quality of their students. In this context, reputation (similar to branding) becomes a key to individual advancement in the labor market. To compete successfully in the war for talent, it is necessary to go to the best universities, or those with the greatest reputational capital. In Brown and Hesketh's (2004) research, it was the Oxbridge type of universities that were at the top of the scale because the employers they based their study on perceived such institutions to have the greatest reputations. However, because students from disadvantaged backgrounds are less able to secure places in elite institutions, they lack access to the reputational capital needed to gain entry into top positions in the labor market.[2] Roizen and Jepson (1985) report similar findings, showing that employers prefer Oxbridge graduates. However, these authors note that "'on the day', the right non-academic qualifications will override A levels, course and institution for very many jobs and that for another substantial proportion of jobs, knowledge and skills are the most important factors" (Roizen and Jepson 1985, 168).

Although it is possible to agree with the sentiment of Brown and Hesketh (2004), that such processes are undoubtedly present in some fields (such as those they investigated), several works have shown that the financial outcomes from succeeding in tertiary education vary widely and in ways that are not always directly attributable to the reputation of institutions, nor to individual departments within providers of tertiary education. For example, in research by Morley (2007), the reputations of institutions and the reputations of departments were ranked by employers in a variety of fields as being among the least important consideration during short-listing and the final selection of new recruits. Morley's evidence is weakened, however, by the diversity of employers who participated in her study. Differences in the kind of degrees taken, differences in the location in which graduates work, differences in graduates' background characteristics, labor market change, and a myriad of other factors all effect graduate outcomes (see Chapter 2).[3] Such factors mean that it remains unclear whether advantage is a result of the students' study (such as the programs taken) or of other factors present before the students began their studies (or, indeed, if it is a result of some combination of these and other unobserved factors). At a general level then, *some* research provides empirical support for the theory that reputation counts (as well as some that suggests otherwise). However, the source of advantage remains unclear. In this respect, Roizen and Jepson (1985) raise a valuable observation; namely, that we do not know how reputational capital might form in different fields.[4]

It is also important to remember that reputation is not merely an abstract category that, for example, reflects one's class position. Rather, it ultimately arises out of repeated interactions, which are beneficial to all parties. In other words, reputation is a form of process-based trust, which is itself a product of repeated exchange. Reputational capital emerges because employers have come to trust that those who graduate from particular institutions are more likely to have the qualities they desire than those from other institutions. Over time, as attending an elite institution may become short-hand for quality, categorical inequality may develop (Tilly 1998), However, it is important to remember that if employers were unable to gain advantage through these interactions, the reputation of providers would ultimately decline (Chevalier and Conlon 2003).[5]

All of this says little about the actual source of reputation (or the source of advantage gained by employers) in the first instance. Indeed, an important task for future research is to understand better the sources of reputation where it exists. This book theorized that in contexts where employers require high levels of technical and scientific expertise, institutions that produce graduates most proficient in required technical and scientific skills would be best placed to develop a positive reputation with employers. According to this account, the formation of reputation is likely to result from flows of positional technical or scientific knowledge. If this proved to be the case, it would help link the formation of reputational capital to actual knowledge produced in research-intensive institutions (and, by implication, link reputation to the quality of instruction).[6] In turn, this would help provide a way to understand more clearly why research-intensive institutions are perceived to be more richly endowed with reputational capital than other providers of tertiary education. In other employment contexts, such as those where employers require aesthetic qualities that are difficult—if not impossible—to teach, institutions that can recruit graduates who already hold such qualities will be better placed to develop a positive reputation with employers. Thus, this book has argued that it is a mistake to assume that the sources of reputational capital are identical in all fields. Indeed, it can be argued that reputational capital is a feature of particular fields, that processes that give rise to its formation are field specific, and that there are likely to be important differences in the value of reputation in different tertiary education markets. In other words, processes of reputation formation will vary across time and space.[7]

Closely related to these points is the observation that conceptions of elite—including how such status is signaled and the resources needed to qualify—are also field-specific factors. In some fields, educational qualifications are highly valued and signal elite status, whereas in others, they are not. For example, a doctorate is virtually a prerequisite to work as an academic (but again, in areas of emerging capacity, as in some disciplines such as architecture, these degrees are not required in New Zealand).[8] In contrast, gaining employment as a real estate agent requires comparatively little, if any, formal education qualifications; yet, on the back of a boom in the housing market, elites in this field currently enjoy very high earnings.[9]

Processes of reproduction involving reputational capital may be present in some fields (Brown and Hesketh 2004; Lauder 2007), but it is not necessarily the case that they are present in others. Indeed, it is important to remember that in some fields, we can reasonably expect reputational capital *not* to offer any explanatory power in understanding graduate outcomes. For example, the rate of graduate earnings is set by bureaucratic wage-fixing methods in many labor market sections, such as nursing and teaching. Both elite and nonelite institutions produce graduates for work in these fields, and although there is more research to be done, it is doubtful that the reputational capital of the institution from which a qualification is gained affects outcomes greatly, particularly earnings. Finally, a further complicating factor is that fields evolve as boundaries shift. For example, during periods where the growth and demand for workers who hold certain skills outstrips supply, employers are forced to look more widely for new recruits. Again, such pressures are likely to affect the importance of reputational capital. For example, current high demand for medical doctors and nurses has forced hospitals in New Zealand to employ virtually all those who hold the required formal qualifications.

This book has argued that one reason we have difficulty properly understanding the relationship between tertiary education and advancement is that researchers (along with others, such as politicians) have tended to draw conclusions about the relationship between participation in tertiary education and social class reproduction across *all* fields. Indeed, most accounts present the view that similar processes of social reproduction are occurring in all areas of the labor market. This means that the presence of reputational capital on outcomes, and the precise mechanisms that generate such capital, remain unobserved or hidden in aggregate quantitative data.

This book has described two interrelated arguments that can be used against the need to look at processes of social reproduction within fields—or at least it can be argued that in the long run, field theory will be of reduced value to researchers who want to understand how the competition for advancement is organized. The first of the arguments is referred to in this book as the technocratic-meritocratic perspective. It has been argued that technological change and the increasing significance of education and training in the labor market (or, more precisely, the increased significance of credentials to advancement) have increased the bond between the tertiary education sector and the labor market across all fields, as more occupations require workers to hold similarly high levels of skill and qualifications.[10] Such processes of isomorphism reduce the value of looking at systems of social reproduction with discrete fields.

This view resonates with the training gospel, or the notion that ongoing investment in up-skilling is needed to meet future skill needs, which is a perspective regularly espoused by recent governments. However, neither labor market change nor employers' behavior provide strong support for this theory.

First, it is common for there to be rapid expansion in many areas of the labor market in which little, if any, demand is made for recruits to have high-level qualifications.

Second, there is little evidence that all qualifications are valued equally across all fields, and that employers use them as a basis for their recruitment decisions. One problem here is that qualifications do not necessarily signal the skills and qualities employers demand. In this respect, recent evidence suggests that firms are increasing their reliance on psychometric testing as a means of selecting new recruits. As John Rust, director of the University of Cambridge's Psychometric Centre, recently put it, "increasingly employers have to use psychometric tests because degree classifications is such a variable quantity these days and they are so broad and in very different subjects. Employers can't tell a graduate's competencies from their degrees" (Lipsett 2007).

Third, processes of innovation mean that returns to individuals from education can be enhanced by the formation of new fields, where capacity is thin and where new rules of advancement are present. New fields can be formed by combining two existing fields. For example, computer science (and the information technology revolution) did not emerge spontaneously from advancements in technology and device manufacturing. Rather, they emerged as a result of three factors: theoretical principles, laboratory research, and business efforts to make computers marketable. In addition, two lines of science had to combine before computers were possible. These were harnessing electricity for data storage and the development of calculus into Boolean logic (Cortada 1993).

Fourth, as argued throughout this book, fields evolve in ways that change the nature of competition for advancement. For example, it is argued that careers in information technology changed substantially in the 1990s with evidence from Europe suggesting that technical interest as a key competency of information technology has been replaced by skills associated with boundary spanning (Loogma et al. 2004). Rather than producing isomorphism, then, field formation (which is driven by processes of innovation and other factors, such as intervention by the state) drives the demand for workers who have new skills.

Fifth, in fields where capacity is thin, the rules of advancement arguably are likely to require workers to have lower levels of formal qualifications (as employers broaden their search strategies and lower the level of qualifications needed to gain entry into positions). In contrast, in fields where capacity is thick, employers can afford to lift the bar well beyond the level needed to fulfil the technical, scientific, and social requirements of the job. In other words, selection processes may well drive up the need for certain forms of capital far beyond the level needed to undertake the tasks at hand. Thus, processes of credential inflation (and deflation) are arguably field specific and vary across time and space.

Finally, changes in economic policy can have a dramatic effect on the evolution of fields and the resources required for advancement within them.

For example, there is anecdotal evidence that Indian film makers are now bypassing New Zealand because they can receive greater financial incentives to produce in other countries, such as Singapore (Field 2007). The wider implications of such changes remains are not known. Nevertheless, it implies that the demand for workers in the field in New Zealand could be reduced.

The second, and related, reason identified in this book for why we might reasonably expect to see uniformity in rules of advancement emerging across all areas of the labor market is the result of government intervention, and particularly interventions intended to create open competitions for advancement. For much of the latter part of the twentieth century, education reform has been geared toward creating meritocratic rules of advancement that tighten the bonds between education and the labor market. In terms of assessment reform, the development of national qualifications frameworks and the associated use of outcomes-based assessment are designed to tighten the bond between education and the labor market by providing more instructive and relevant information to users of qualifications. Frameworks have also provided a useful way of furthering the attempt to create global currencies in education. By providing better-quality information, frameworks are designed to improve transparency and openness in education and training systems and to encourage employers' use of bureaucratic (and, hence, meritocratic) recruitment methods. The clearer specification of the competencies demanded by employers is also thought to allow learners to make more informed judgements about just what training they need to advance in the labor market. This is understood to reduce the importance of knowing who by transferring knowledge that was formerly embedded in social networks—and that remained closed to outsiders—into new educational technologies (e.g., in the forms of new curricula and new standards of achievement). By shifting ownership of skill away from, for example, those who participate in closed learning communities, and toward all learners, frameworks are thought to democratize access to knowledge and thereby promote openness in access to innovative knowledge. However, evidence suggests that this view is not supported by either labor market change or the behaviors of employers (who tend not to value framework qualifications as highly as predicted). In addition, recent changes in government policy are designed to transform tertiary education by increasing the differences between providers of tertiary education.

A further reason we can discount any tendency toward isomorphism is that access to innovative knowledge is not necessarily characterized by openness. Rather, access to many forms of innovative knowledge is positional in nature. For example, as Zucker and Darby's (1996) research on the formation of the U.S. biotechnology sector shows, the knowledge created by star scientists initially provided competitive advantage. However, eventually this knowledge spread out into other universities and became routinely taught. This increased access to knowledge needed in the field; indeed, it arguably democratized this knowledge. However, another consequence of increasing the access to this knowledge was that its value as a means of individuals gaining competitive advantage declined. Thus, democratizing some forms of knowledge is associated with reduced economic returns to those who hold this knowledge. Indeed, as argued throughout this book, such knowledge ceases to have positional qualities.[11]

Although "third way" governments remain committed in theory to the idea of equality of opportunity through education, in practice, network creation is cutting against this goal. Although innovation is closely bound up with networks, to date, researchers have not properly considered how intervention by government helps create contrasting geographies of talent, nor have they considered resulting social implications. Network formation makes economic sense, but social advantages are not as evident. Indeed, on the basis of the analysis presented in this book, network formation and social inclusion appear to be incompatible objectives.

In this respect, there are at least three good reasons why networks will play an increasing role in competition for advancement through tertiary education. First, networks provide both a conduit through which innovative knowledge flows and ways to create new or innovative knowledge. However, participation in such knowledge flow is limited to insiders. Second, the skills and qualities demanded by employers appear to be poorly represented by qualifications achieved. As a result, employers rely on other, field-specific, signals of competency, such as an individual's reputation in the field or their charisma (Brown and Hesketh 2004). Networks provide a reliable and trustworthy source of information about the potential of new recruits, particularly where other proxies of the qualities and skills employers desire are lacking. Third, although third way governments continue to advance a discourse of social inclusion and are promoting educational reforms designed to assist in the formation of open competitions for advancement, much emphasis is being placed on network formation as a means of increasing innovation. For these reasons, the general argument of this book is that we could reasonably expect that network processes would help account for differences in graduate outcomes in specific fields. For example, it was argued that those who train in institutions where innovative knowledge is created are likely to be better placed to win a position in fields where competitive advantage is derived through access to positional knowledge. This chapter returns to these issues later. For the moment, attention can be turned to assessing what light, if any, the empirical chapters can shed on our understanding of the issues.

Evidence from the Field

The purpose of the case studies was to use the testimony of managers who employ graduates to improve our understanding of competition for advancement through tertiary education. It was argued that we needed to look within fields to understand how the rules of advancement operate and to understand how reputation develops. In general, the research conducted as part of this book allows us to draw conclusions in three broad and interrelated areas: networks and employment, sources of innovative knowledge, and employers' links with tertiary education.

Networks and Employment

As Granovetter (1995) showed, despite economic changes and modernization of the economy, the importance of networks in the recruitment process continues. The data gathered in the New Zealand–based case studies supports Granovetter's findings. In general, managers reported a heavy reliance on social networks to recruit workers in both the screen production and biotechnology sectors. Networks provided an effective conduit through which reliable information about the capacity of individuals was transmitted. However, there were variances in the way they were used by managers in the two New Zealand cases.

In the biotechnology sector, higher-level qualifications are necessary for advancement, as employers require evidence that potential recruits have the scientific and technological skills needed. However, qualifications are not enough to guarantee an individual's progression. Instead, individuals needed to augment qualifications with network resources (and, by implication, possess the cultural and social skills needed to create and use networks). The importance of network resources was encouraged in the biotechnology sector by two factors.

First, it was not just that the kinds of skill needed could not be accurately signaled through qualifications, it was also the case that skill demands needed in biotechnology firms went beyond those that could reasonably be produced in tertiary education. In this respect, it was not the case that students in certain elite institutions had access to skills not available to those who studied elsewhere (as theorized above), or that employers favored graduates from these institutions ahead of others. Rather, the managers interviewed for this book were more interested in several key factors: the depth of understanding potential recruits had in their areas of expertise (which employers perceived could be initially developed at any university with the appropriate programs), whether or not they had the ability to use this understanding to produce commercial products (i.e., whether or not they possessed the required tacit skills, such as the ability to be entrepreneurial), and their broader work experience. Thus, no particular tertiary institution was dominating the provision of the skills needed by the firms. Rather, graduates needed to demonstrate their own development of the technical capacity needed to advance the products firms were creating. At this juncture, it is worth restating three points.

First, the technologies being developed by these firms are cutting edge, and their geographies of talent are global. In part, this reflects the fact that, in this field, advancing technologies requires gaining access to the widest pool of talent. Accessing this pool of talent requires global recruitment strategies and frequently involves the use of international network recruitment strategies. Universities participate in these networks by providing consultancy services and the initial intellectual property that was required, and also by training individuals in the skills needed. However, in the testimony of interviewed managers, no hierarchy of institutions was detected.

Second, because of the emergent stage of New Zealand's biotechnology field, domestically trained workers lacked the required skills and experience needed to advance the technologies being developed by the firms. Thus, early in the development processes, the skills needed to advance biotechnology products were only available internationally. For example, overseas workers already had the skills and depth of experience needed to help firms meet the regulatory requirements for product registration and, hence, to realize their commercial potential.[12]

It may be that in time, as local capacity increases, reputational capital will develop. For example, further advancement of the biological cluster by developing a new biotechnology center at the University of Auckland may have an important impact on the field. However, at present, reputation is best seen as an attribute of individuals, rather than as a result of where individuals studied. The reputations of scientific and technical recruits are, therefore, created through the particular work they have undertaken (which, in part, was seen to reflect work individuals had undertaken during their studies), the research they had published in peer-reviewed journals, and the networks they had formed.

In the screen production field, the data also show that networks are critical to recruitment. One reason for this is that innovation was primarily derived through the development of cultural products that are attractive to audiences. In part, innovation involved the identification of suitable topics for new programs and their production for markets. However, the ability to produce creative works for commercial markets was not signaled by qualifications. Expertise was signaled by the creative works potential recruits had been involved in, coupled with information about their capacity that was transmitted through social networks. Finally, it is worth noting that the field is not one in which graduates tend to enter elite positions in the first instance. Rather, all start off at the bottom of the industry and must prove themselves at every level.

In the design case study, the procedure the managers used to recruit new staff was different from the case studies reviewed above. One difference was the use of advertisement and interview as key recruitment strategies, which were mandated by the state (as a major funder). In some instances, particularly those where high levels of technical and scientific knowledge were required and capacity in the area was thin, the managers knew of potential recruits, either directly or indirectly (implying the presence of network processes). In such cases, the managers reported that the skills needed could only be obtained from very few research institutions. However, where the technical and scientific knowledge was widely available and capacity in the areas was large, in winning a position, knowing who was of lesser importance than knowing what.

In several examples, the reputation of knowledge partners and institutions they were employed in remained important to the establishment of knowledge transfer partnerships. In this respect, differences between providers of tertiary education, in terms of their ability to provide innovative knowledge, were presented within design knowledge transfer partnerships canvassed as part of this research. For example, in protein manipulation, Thompson's University was identified as a reputable provider of the required skill and expertise needed to solve one firm's design problems. However, circumstances of time and place and other activities (such as cold calling on firms) also provided a way for knowledge transfer partnerships to develop.

Sources of Innovative Knowledge and Links with Providers of Tertiary Education

The links between providers of tertiary education and innovative firms are clear in Auckland's biotechnology sector, and also in the design knowledge transfer partnerships in England, but not in the creative field in Wellington. In the biotechnology sector, formation of linkages was driven initially by star scientists many of whom worked in institutions of tertiary education who drew on their social networks to advance their innovation. As shown earlier, very soon into the innovation process, the biotechnology firms where the informants worked sought new recruits through exploiting their global networks. Their participation in global networks is underpinned by the need to gain access to the most innovative forms of knowledge and by a lack of local capacity.

In biotechnology, it was important that new recruits held scientific skills, or the vocabulary required to be effective in the workplace. These could only be gained from universities that provided postgraduate training in research. The quality of training was also important, but higher-level qualifications did not always signal this. For example, managers reported that some providers of tertiary education operated doctorate factories, producing graduates *en masse*. This raised questions about the quality of the programs on offer.

Although the linkages between biotechnology firms and particular providers of tertiary education were apparent at a senior level, such as where some chief scientific officers were also employed as senior academics at Auckland University, in general, managers used global networks in the recruitment of scientific staff. It did not appear to matter where new recruits had gained their qualifications—so long as they could demonstrate technical expertise, they had met a key criterion. As stated earlier, capacity in highly specialized areas was thin, and most potential recruits could be identified and assessed through social networks. These networks were necessarily global. Indeed, knowledge created in other countries was typically needed to advance products being developed locally. Thus, the geography of talent was necessarily global, and firms needed to exploit this if they were to advance their products. Indeed, global ecologies of skill in which firms needed to participate meant that local providers operated at a distance to local firms and did not seem to provide a conduit through which their graduates could flow.

In the screen production field, knowledge considered innovative included creative ideas for productions and knowledge of the sorts of projects funding bodies would likely support. In both instances, these factors were sourced through the managers' social networks in which providers of tertiary education did not directly participate. There was no evidence that providers of tertiary education were producing innovative knowledge of a kind that could be used directly by employers in the field.[13] There were few (if any) explicit linkages between firms and providers of tertiary education in the screen production field. In this field, competitive advantage was derived through the production of creative works for audiences in this field. In general, employers did not report using educational qualifications as a basis for their recruitment decisions (though there were exceptions; e.g., researchers needed particular qualifications). To be sure, new recruits might well hold these, but they were not actively sought out as signals of capacity. Instead, other indicators of capacity were relied on, including those with whom potential recruits had worked in the past and those in the industry who could recommend them. An important reason for this, identified by the interviewees, rested in the fact that working effectively in the field required individuals to possess creative, social, and cultural skills that were difficult, if not impossible, to teach. For example, being effective on set required that all workers be team players. Furthermore, far from emphasizing creativity, much of the work on offer required that workers remain disciplined and keep their heads down.

In contrast to the New Zealand cases, the design knowledge transfer partnerships links created between firms and the knowledge bases were vital to innovation. Indeed, in the knowledge transfer partnerships, links with providers of tertiary education were critical to developing innovative solutions to technical problems. However, in contrast to claims advanced earlier in this book, in general, participation in partnerships was not dominated by elite institutions. Indeed, official statistics reveal that lower-ranked institutions were more likely to participate in the program than those ranked highly in the Research Assessment Exercise. Finally, in areas where the knowledge needed to develop the innovation was highly specialized and technical in nature, there were relatively few providers of tertiary education who had the required skills and expertise.

Implications for Researchers and Policy Makers

The presence of different rules of advancement in different fields has implications for policy makers and for researchers. Until recently in New Zealand, policy interventions (and the associated educational discourse) have tended to treat all qualifications as being of equal value in the labor market. As described above, the clearest expression of this is in the training gospel, or the notion promoted by officials that economic and social changes are driving up demand for highly skilled and highly qualified workers (and, hence, increasing the overall need for education and training). Politicians have also been attracted to such ideas because it has allowed them to hold out the promise of social mobility through tertiary education. The work this does in terms of legitimating inequality has long been acknowledged by sociologists of education (Brown and Lauder 2001). However, the policy is based on a number of assumptions, which are open to question.

First, it assumes that the emergence of the knowledge economy and developments in education would encourage isomorphism. As noted above, this is not the case. In some fields, educational qualifications are highly valued, whereas in others they are not. Indeed, fields differ considerably in terms of the mix of resources needed for advancement, and the associated values of these vary over time and space. For example, in some fields, the use of open recruitment processes is standard. However, in others, such as those in which capacity is thin or the skills required highly specialized, network recruitment was most effective. In any case, on the small amount of evidence presented above, even where open recruitment methods are used to recruit workers in thin fields, the small size of the pool markedly increases the likelihood that many applicants will already be known to recruiters. Over the long haul, isomorphism is incompatible with innovation.

Second, it assumes that the graduate labor market is increasingly becoming the norm. For this reason, policies should be geared toward increasing the proportion of people in tertiary education and training. Again, this assumption is highly questionable. In some fields, graduates do not appear to have any advantage over nongraduates. In the creative field, for example, although some specialist positions are only available to tertiary-trained graduates (e.g., researchers), in many areas of the creative labour market, all workers, irrespective of their training, start off at the bottom and demonstrate their capacity through the creative works they help produce, as well as their demonstrated social skills.

Third, it assumes an ability of qualifications to signal capacity. The training gospel maintains that educational reform is bringing the qualifications produced in education and training systems into closer association with the labor market. This perspective emphasizes increasing demands from employers for new recruits to hold high levels of skill and expertise, along with qualifications supposedly signaling these attributes. However, as Brown and Hesketh (2004) make clear, credential inflation and expansion of tertiary education have reduced the ability of qualifications to signal capacity. Providers of tertiary education have attempted to increase the ability of qualifications to signal capacity by such moves as development of ever more specialist qualifications and the implementation of outcomes-based assessment. However, increasing use of psychometric testing suggests that these and other strategies designed to increase the signaling capacity of educational qualifications have failed to provide a sufficiently high quality of information to employers. Encouraged by their political masters, providers of tertiary education have further compounded the problem by producing many more graduates than can reasonably be absorbed by the labor market. As a result, employers are seeking out new ways of distinguishing between graduates while maintaining their use of traditional methods, including employing network recruitment methods and using psychometric tests.

Fourth, it assumes the existence of a strong and direct connection between innovation and incomes. The discourse underpinning the knowledge economy is largely maintained by the argument that competitiveness is derived through technological and scientific innovation. Although competitive advantage is related to innovation, it is not clear that certain fields must themselves be innovative to benefit from innovations made in other fields.[14] What this means is that certain fields can remain insulated from competitive pressures (e.g., by instituting forms of social closure). In a similar refrain, as Weber noted many years ago (Murphy 1984), certain fields maintain the incomes of incumbents through enacting forms of closure that restrict membership. In this context, competitive advantage is not derived through innovative knowledge. Rather, it is derived through strategies that insulate individuals from market pressures. Such practices also mean that recruitment practices in these fields can remain moribund. In such instances, then, increased earnings may have little relationship to changes in the work that is being undertaken or to any increase in the knowledge needed to be effective in the field. Furthermore, recruitment can be both class based and, as Kanter (cited in DiMaggio and Powell 1983) put it, homosexual.

These comments give rise to a number of implications. If lifting the participation of those people from disadvantaged backgrounds is designed to increase their competitive ability for advancement through tertiary education, greater attention needs to be placed on the actual kind of participation. As researchers have pointed out, in the case of the United Kingdom, just about all of those who are qualified to attend higher education do so, and there is no evidence of large-scale inequity in admissions to higher education (Gorard 2005).[15] However, on the basis of this study, it seems that increased retention and attainment (with a view to increasing participation in higher education) is only likely to pay dividends if the rules of competition for advancement in fields students eventually compete in actually involves the use of qualifications. More pointedly, by encouraging students into programs that have little or no relevance to labor market policies, promoting participation may do more harm than good. It is worth noting that employers, providers of tertiary education, and students do not compete in the same way, nor do they compete in the same markets. One obvious illustration of this is that in general, the market for students is much larger than the market for graduates. Similarly, providers of tertiary education have been encouraged by government and by internal forces to produce more graduates than can be accommodated in the labor market. In this respect, policies that encourage the concentration of research capacity, along with the production of postgraduate students in particular institutions, do not in themselves help build an institution's reputation. For example, simply completing a doctorate does not in itself signal an ability to be effective in biotechnology. Rather, it is the broader qualities of the graduates produced by institutions of tertiary education that count in terms of building a reputation with employers.

For students, the advice of this book would be to consider the rules of advancement in operation in the field in which you hope to work. Contrary to Brown and Hesketh (2004), it seems unlikely that the cost of attending an expensive elite institution is worth the investment when the rules of advancement in one's chosen field do not require qualifications from such institutions.[16] In this context, it is worth repeating James et al.'s (1989) argument that although sending one's child to Harvard appeared to be a good financial investment, sending them to a local state university to major in engineering and encouraging them to gain a good grade point average was an even better investment.

For providers of tertiary education, one message of this book is that building a reputation will require focus on the rules of advancement in fields they serve.

The Final Word

The basic message in this book is that processes of social reproduction (including processes of reputation formation) vary across time and space. As a result, all involved in tertiary education would gain from a close inspection of the rules of advancement in operation in fields in which they are involved. Furthermore, the forms of capital needed to advance, including reputational capital, are best understood in terms of the geographies of talent that characterize these fields.

Field theory provides a way to identify more clearly the mix of resources needed to advance. However, although field theory provides a way to isolate the precise resources needed for advancement and allows for the easier identification of the rules of advancement present, operationalizing the notion of field in the manner outlined above has limitations. In this respect, the data show that fields are more diverse than implied by the theory, with managers reporting the use of different strategies for recruiting staff in different contexts.

This observation also holds relevance for our understanding of systems of innovation. In this respect, field theory alerts us to the presence of systems of innovation, which are nested in different ways, there being both national and international systems of innovation. The contribution made by skilled labor (in its various forms) to innovation, and the relationship between this and tertiary education providers, are field specific. Thus, there is no one system of innovation, with uniform rules of advancement that might guide policy makers interested in building high-wage, high-skill economies.[17]

The purpose of this book has been to explore the relationships among innovation, social networks, and the competition for advancement through tertiary education. It has aimed to contribute to debates by describing how network formation affects this competition and by describing how educational policies are attempts to form, reform, and erase boundaries between fields. By advancing field theory as a way of providing a better conception of how competition is organized, this book provides a way for researchers and policy makers to understand how inequality is transmitted across generations and to understand the effect of the state on the functioning of fields. If it has provided a way to deepen our knowledge about how the competition for advancement through tertiary education operates, it will have achieved its purpose.

Notes

Chapter 1

1. The Research Assessment Excercise and the Performance-Based Research Fund are both designed to fund those institutions which have the most research active staff. Although there are important differences between the two policies, in general, they achieve this by requiring all institutions who would like funding to undertake research to submit details of their staffs research outputs every seven or so years.

2. Brown and Hesketh (2004) interviewed graduates from 15 "leading edge" private and public sector companies. All of the private sector companies were multinationals offering the possiblility of fast-track appointments and international placements. Organizations were selected to give a comprehensive picture across a range of business sectors that included telecommunications, financial services, manufacturing, and pharmaceuticals.

3. Policy makers have also tended to assume that the relationships among education, training, and the labor market are uniform. For example, even though research suggests otherwise, policy makers tend to assume that educational qualifications are needed for entry into the labor market and have invested heavily in their production.

4. Exceptions to this claim can include forms of knowledge that become more valuable as the number of people who use it increases. For example, the value of the Internet has increased as the number of users has increased.

5. Brown and Hesketh (2004) argue that employers are responding to such challenges through recruiting from reputable institutions, in part because this reduces the cost of recruitment.

6. "Third way" administrations are those that draw on both neoliberal and social democratic methods of social and economic administration (Giddens 1998).

Chapter 2

1. Organizations in Brown and Hesketh's (2004) research were selected to give a comprehensive picture of the practices across a range of business sectors, such as retail, financial services, telecommunications, manufacturing, and pharmaceuticals, and included fifteen leading-edge private and public sector organizations. All of the private sector companies were multinationals that offered the possibility of international assignments and training opportunities for their fast-track appointees. Six of these organizations were selected as case studies, including two from the public sector.

2. Another study from the United Kingdom has shown that male university graduates earn a wage 48 percent higher, and female university graduates 43 percent higher, than those without formal qualifications (Conlon and Chevalier 2002).

3. Similarly, although competition between nations remains an indelible feature of advanced capitalist societies, nation states need to work collectively to win in the global competitions for advancement. For example, in the United Kingdom, Tony Blair has advanced a discourse of regionalism in which making Europe strong is critical to keeping pace in an increasingly globalized world.

4. However, when considering such critiques, it is important to remember that the point of the technocratic-meritocratic perspective is not that it represents the truth but, rather, that it helps legitimate interventions the state wants to make.

5. See Levine (1999) and Johnson and Reed (1996) for British research, Solon (1992) for research carried out in the United States, and Whelan and Layte (2002) for Irish-based evidence.

6. The association between origins and destinations has led some to argue that effort should be placed on increasing the foundation skills of children before they reach tertiary education (Corak 2004). Although this position has merit, it does not mean that tertiary education is not part of the problem. Students from low socioeconomic backgrounds could simply see insurmountable barriers to entering tertiary education and exert less effort in their schooling as a result. It is also likely that class-based processes affect how individuals select what institutions of higher education they attend (Ball 2003).

7. Blundell et al. (2005) found in their study of the National Child Development Survey in the United Kingdom that in general, there is a great deal of variation in the returns on education across individuals who held the same educational qualifications.

8. The Russell Group is an association of 20 major research-intensive universities of the United Kingdom.

9. Dale and Kruger (2002) also found, however, that children from low-income backgrounds earned more if they attended selective colleges.

10. In many respects, the problems faced by researchers in the area are similar to those faced by those working in school effectiveness research. In this branch of sociology of education, researchers are debating the extent to which schools impact on student achievement (Teddlie and Reynolds 2001).

11. For example, although Carole Middleton was a self-made millionaire, she could not ensure that her daughter, Kate, married into royalty by marrying Prince William. It was reported that Carol was "pushy, rather twee and incredibly middle-class and used words such as 'Pleased to meet you', 'toilet' and 'pardon'. In addition, Kate felt isolated and abandoned at realising she would never be accepted by William's aristocratic friends, nor the Queen Mother" (Allen 2007).

12. However, note the debate about the different kinds of culture needed in different academic fields that was briefly outlined in the Introduction to this book.

13. Tilly argues convincingly that inequality forms around processes or ongoing transactions, not on the basis of the preexisting attributes of immigrant groups. Thus, power cannot be understood as an individual attribute—it is embedded in transactions and interactions between individuals. These interactions also help construct individual identity. In turn, these identities give rise to the formation of obligations and expectations and to the formation of new networks. Ultimately, networks become categories and so on to the formation of groups. Once categories become institutionalized within organizations (construed broadly to include well-bounded clusters of social relations, such as firms, kin groups, and local communities), they can be considered durable. For Tilly, the usual effect of the formation of categories is social exclusion, or the exclusion of one group of participants in a network from resources controlled by participants in another network.

14. This approach encourages capitals to proliferate. For example, emotional capital has recently entered educational debates (Reay 2000).

15. Unfortunately, there has been relatively little empirical research exploring the relationship between different capitals operating in different settings. One difficulty is that it has proven difficult to isolate the cause of observed effects. For example, in their study of American citizenship and political power, Nie and colleagues (1996) argue that there exist a limited number of positions of power in any network, and educational attainment is the best predictor of network centrality. However, as Egerton (2002) points out, these results might actually confound educational achievement with the social origins. Indeed, Egerton detected differences in the nature of the civic networks used by professional and managerial groups. Generalizing these results to the wider economy, it can be argued that the natures of the networks needed for advancement in banking, government, technological clusters, and local labor markets differ.

16. Defining a field as a specific occupation offers challenges too, as within-occupation income inequality is greater than that that found between occupations, at least in the United States (Weeden et al. 2005).

17. As Marx insisted, capitalism will never survive unless the social conditions that support it are also reproduced (Cutler et al. 1977). This means that individual subjectivities, or identities, such as those that relate to work, consumption, and inequality, need to support the productive forces at play. In this respect, any modification to the rules of advancement must include both changes in the productive forces at play *and* changes to the subjectivities that support these.

Chapter 3

1. For example, the emergence of boundary-less careers, in which careers are increasingly characterized by interfirm mobility, has raised the possibility of intrafield mobility as an emerging trend.

2. In other evidence, researchers have identified a beauty effect in which those perceived to be more beautiful earn a premium of approximately 10 to 15 percent over their plainer peers (Hamermesh and Biddle 1994).

3. For example, there is some research that suggests that although actually having friends in high places appears to have little effect, those who are perceived as having friends in high places enjoy enhanced status themselves. For such reasons, being seen by others to participate in the right networks is important to gaining mobility (Kilduff and Krackhardt 1994).

4. When considering these comments, it is important to remember that the style labor market is diverse; different kinds of aesthetic labor are needed in different contexts. For example, the kinds of aesthetic labor power needed to be effective in a boutique hotel will differ from those needed to sell insurance or assist passengers on an airplane.

5. However, such investment also helps establish new forms of social closure, such as by limiting access to valuable information to insiders, but, as Cranford (2005) notes in the case of the janitorial industry in Los Angeles, network recruitment can be exploitative—and can serve to reduce social mobility—by limiting exposure to new information and opportunities.

6. Understanding network formation as both an economic and a cultural process is useful, as it highlights the difficulty faced in gaining access to sources of competitive advantage that are embedded in culturally alien networks. A critical link between Silicon Valley's fifty or so research centers and Stanford University is the informal and decentralized forums, which make it easier for foreign companies to learn the Silicon Valley way, or to introduce themselves to its sometimes-arcane culture (Casteilla et al. 2000).

7. From a left-wing perspective, such actions are attempts to reduce transaction costs by limiting the ability of workers to retain ownership of skill (Jordan and Strathdee 2001).

8. This point is open to the criticism that competitive advantage will increase as a result increasing the number of people who hold qualification. To date, in part a lack of a critical mass of qualified workers has damaged competitiveness by reducing the attractiveness of nation states as a place to invest in high-skill, high-wage employment (Brown and Lauder 2001).

9. Although dated, there is some evidence from Japan that the creation of internal systems of promotion with firms, which encourage the development of lifetime careers, is designed to reduce this problem and to increase the incentives for employers to invest in the training of core workers (Sako 1999).

10. Hayek's (1945) stress on the advantages of markets in this context remains disputed by those who argue that, "to run a firm as if it were a set of markets, is

ill-founded. Firms replace markets when non-market means of coordination and commitment are superior" (Rumelt et al. 1991, 19). The key point of relevance to the present discussion is that access to unorganized knowledge and knowledge of circumstances of time and place offer competitive advantage.

11. Morley's (2007) research is based on a diverse range of employers, and the likelihood that the rules of advancement are different in different fields is not considered.

Chapter 4

1. This chapter is a revised and extended version of Strathdee (2006).
2. Although the account presented here is that the creation of open and closed geographies of talent are contrasting and in conflict, some researchers present network creation and the development of open geographies of talent as being compatible with one another (Szreter 1998).
3. It could also be argued that, by constructing networks, such as those emerging out of the formation of knowledge transfer partnerships, the state is actually undermining advantage that elites derive from their own closed networks. In turn, a case that the construction of networks by the state has the ability to democratize the social infrastructure can be mounted.
4. It is worth noting that there exists a distinct moral element in third way attempts to repair deficits in the social infrastructure. The emphasis is on distancing individuals from the relationships and networks that might limit their participation in employment or education and training (Strathdee 2005b).
5. Some providers in areas deemed by the government to be at risk of performing poorly have been given shorter funding guarantees.
6. See Chapter 6 for discussion.
7. Of course it is more complicated than this. In the new environment, the government will make the final decisions about what courses it funds. However, it is relying on a variety of actors (e.g., employers, Maori, and other interested individuals groups) to develop a shared understanding of stakeholder needs on a national and regional basis. By bringing together all those with an interest in skill development together to pool information and gain a shared view of tertiary education needs, the government hopes that priorities and gaps in tertiary education can be identified.
8. The key objectives of economic transformation agenda include building a shared understanding of regional tertiary education needs, gaps, and priorities; developing system capability at a regional level through collaborative relationships; better outcomes for learners and other stakeholders; and efficiency benefits and capability development for tertiary providers and stakeholders (Tertiary Education Commission 2007).
9. It is important to note, however, that both New Labour and the Labour-led Coalition have sought to protect the status of academic routes into selective tertiary programs.

10. In this context, process-based trust, or reputation, becomes significant. A key reason for this is that social capital of the kind that promotes innovation is enhanced through repeated exchange, which gives rise to reputation. In addition, there are important cultural and linguistic aspects to knowledge transfer, with shared language and cultural codes enhancing innovative potential.

11. See Chapter 7 for further detail.

Chapter 5

1. The New Zealand Screen Council is an independent industry organization with a responsibility to facilitate growth within the New Zealand screen production sector (http://www.nzscreencouncil.co.nz/home.shtml).

Chapter 6

1. In the 2006 round, researchers who had become Performance-Based Research Fund eligible since the first round were required to submit portfolios. For example, those researchers who were appointed after the first round were required to submit portfolios for assessment. All others had the option to submit or not.

2. Because students are required to pay to access knowledge, the introduction of tuition fees is another example of this process.

3. These are the Allan Wilson Centre for Molecular Ecology and Evolution (based at Massey University, with partners from the University of Canterbury, University of Auckland, University of Otago, and Victoria University of Wellington), the Maurice Wilkins Centre for Molecular Biodiscovery (at the University of Auckland), the MacDiarmid Institute for Advanced Materials and Nanotechnology (at Victoria University of Wellington, with partners at the University of Canterbury, Industrial Research Limited, and Institute of Geological and Nuclear Sciences), the National Centre for Advanced Bio-Protection Technologies (at Lincoln University, with partners at Massey University, New Zealand Crop and Food Research Ltd., and AgResearch Ltd.), the National Centre for Growth and Development (at the University of Auckland, with partners at Massey University and the University of Otago and contributions from AgResearch Ltd.), the New Zealand Institute of Mathematics and its Applications (at the University of Auckland and partnered by New Zealand Mathematics Research Institute [New Zealand MRI] Inc.), and the Nga Pae o te Maramatanga (Horizons of Insight)—the National Institute of Research Excellence for Maori Development and Advancement (at the University of Auckland and partnered by Te Whare Wananga O Awanuiarangi, Te Wananga O Aotearoa, Victoria University of Wellington, University of Otago, University of Waikato, and Landcare Research).

Chapter 7

1. As was the experience with previous chapters, defining the field in this way is likely to raise problems, and it may well be the case that competition for advancement within design knowledge transfer partnerships varies markedly.

2. In turn, the Company Teaching Scheme evolved out of Innovation and Growth Teams and Faradays. Faradays were named after the nineteenth-century scientist Michael Faraday, who maintained strong links with industry while engaged in his experimental work in electricity and chemistry.

3. Knowledge Transfer Partnerships are one of two major products in the Department of Trade and Industry's technology program. The other is Collaborative Research and Development, which is designed help cut the cost of bringing the results of research to market. They aim to achieve this by assisting industry and research communities in working collaboratively in areas deemed to be of strategic importance (science, engineering, and technology).

4. Unfortunately, no information on the relationship between the area in which the knowledge transfer partnership operates and the ratings achieved by the knowledge bases in the Research Assessment Exercise is available.

5. Shenkar and Yuchtman-Yaar (1997) identify self-aggrandizement, or the tendency for individuals to judge their own organizations more favorably than others, as a potential problem.

Chapter 8

1. The Times' League Table uses a range of ranking criteria that cover peer review, research impact (citations per faculty member), faculty-to-student ratio, and international orientation.

2. For such reasons, it is widely argued that processes of social reproduction found in tertiary education (and postgraduate education in particular) given an advantage to elites (Naidoo and Jamieson 2005; Strathdee 2005a).

3. At this juncture, it is useful to restate points made earlier in this book; namely, that selection into tertiary institutions is not random and that it is difficult, if not impossible, to account for all of the factors (even those that can be observed with relative ease) that affect graduate outcomes.

4. In any case, large-scale studies that have attempted to isolate institutional effects on graduate earnings can only identify a relatively small advantage overall to graduates from their attending elite institutions. It is for such reasons that Smith et al. (2000) advise prospective students not to follow performance rankings slavishly. Although the picture is complex and changing, there is some quantitative evidence suggesting that graduates from elite universities do earn more than those from other tertiary institutions, as predicted. For example, Chevalier and Conlon (2003) attribute an additional 0–6 percent for males and a 2.5 percent gain in earnings for females who attend the elite, Russell Group universities. Nevertheless, Chevalier and Conlon conclude that, after accounting for the

observable characteristics of degree subject, academic background, and family background, the quality claim has been largely overstated. However, it is important to stress that the debate continues. For example, Black and Smith (2006) argue that existing accounts underestimate the earnings effects of college quality in the United States, whereas Eliasson (2006) finds, in contrast to most previous Swedish research, no systematic differences in estimated earnings between different colleges.

5. For this reason, Chevalier and Conlon (2003) argue that the advantage they identified in their study offered to those who attended elite institutions derived from human capital rather than network capital. If it were otherwise, the benefit would decline over time.

6. If employers find that graduates lack required skills or qualities, it will be damaging to the reputation of the institution from which they graduated.

7. In addition, there is argument and some evidence that differences in the way tertiary education is structured in various nations actually affects the formation of reputational capital. For example, although their data are dated, Black and colleagues (2005) argue that the differences in college quality they found in the United States may result from the nature of the higher education sector there, which is highly competitive, and may not be present in nations with more highly centralized systems of higher education.

8. Similarly, as described earlier in this book, the kinds of cultural skills needed to gain employment also vary between disciplines. To reiterate, knowledge structures found in the sciences differ from those found in the humanities, raising the possibility that the kinds of capital needed for advancement within the field of higher education differ from discipline to discipline (Snow 1964).

9. A related problem for those who see reputational capital as being of increased importance in the contemporary period is that sustaining their account in fields where the system of innovation is global in nature requires that the reputational capital of elite institutions have global currency. This may be the case, but it remains an empirical question that awaits further research and testing.

10. Closely related to this is the adoption of similar recruitment strategies throughout the labor market. For example, Lauder (1997) argues that small-to-medium enterprises in the United Kingdom are adopting strategies similar to those adopted by the larger multinationals. In other words, they too are targeting the graduates of elite institutions.

11. Social capital theorists might well argue that sharing the knowledge freely will increase innovative capacity by, for example, helping to create open learning communities.

12. The firms found it difficult to compete on price for global talent, though accessing the best talent (if not the only talent) required a global focus. Indeed, employers reported that they did not try and compete on price. Not only were they limited by the financial resources at their disposal (when compared with overseas firms) but they were not prepared to put the whole remuneration structure at risk by paying particular workers far more than their peers. Instead, they

employed other strategies to encourage people to come to New Zealand, including the lifestyle advantages of working there.

13. Though as noted above, this knowledge could be flowing in indirect ways.

14. For example, in countries such as New Zealand and Australia, builders have enjoyed strong increases in income on the back of a buoyant property market, yet there is little evidence that skill demand in the field has increased. If anything, the anecdotal evidence suggests the opposite, with labor shortages leading to a reduction in qualifications and training needed to secure a position in the field. Similar observations can be made in other areas. In a number of different fields, policy makers have reduced the length of time it takes to gain qualifications. For example, in teaching, providers must now offer a three-year Bachelor of Education degree, rather than the four-year qualification that was previously standardized.

15. Given this, these researchers have argued that attention needs to be focussed on addressing why students lack the required qualifications in the first place. In light of this, Gorard (2005) suggests effort should be exerted at addressing such things as differences in retention rates and attainment at the sixth-form level and increasing access to preschool education.

16. In a similar refrain, if gaining a first-class degree is essential, it may be better for students to attend institutions where they will shine, rather than one where they might perform at a level well below their peers.

17. Indeed, there may be tensions between various subsystems of innovation. For example, increasing consumer choice in education might well have increased innovation in some areas (by driving up the provision of skill) but reduced it in others (by delivering training that is not related to the field in which graduates work). All of this suggests that sources of innovation must be identified on a case-by-case basis. In this respect, it remains unclear where New Zealand derives its competitive advantage in the creative sector. In part, New Zealand's film sector has grown because of the influence of a domestic star creative who wanted to be based in New Zealand. However, it is also clear that New Zealand is competing in a global competition to provide production incentives. In this respect, the recent announcement that Bollywood—the world's biggest movie maker—would no longer produce in New Zealand because of the country's refusal to offer sufficient financial incentives is illustrative.

References

Adams, S., and V. Henson-Apollonio. 2002. *Defensive publishing: A strategy for maintaining intellectual property as public goods.* The Hague: International Service for National Agricultural Research.

Admissions to Higher Education Review. 2005. *Fair admissions to higher education: Recommendations for good practice.* Nottingham: Department for Education and Skills.

Adnett, N., and K. Slack. 2007. Are there economic incentives for non-traditional students to enter HE? The labour market as a barrier to widening participation. *Higher Education Quarterly* 61 (1): 23–36.

Agbetsiafa, D. K. 1998. Financial intermediation under information asymmetry: Implications for capital market efficiency in selected developing countries. *Managerial Finance* 24 (3): 62–73.

Ahier, J., and R. Moore. 1999. Post-16 education, semi-dependent youth and the privatisation of Inter-age transfers: Re-theorising youth transition. *British Journal of Sociology of Education* 20 (4): 515–30.

Allen, V. 2007. Snobs who did for Kate: His mates sneered at her mother. *The Mirror*, April 16, 2007.

Amin, A. 2004. Regions unbound: Towards a new politics of place. *Geografiska Annaler: Series B, Human Geography* 86 (1): 33–44.

Amin, A., and N. Thrift. 1994. *Globalization, institutions and regional development in Europe.* Oxford: Oxford University Press.

Andersen, E. 1992. Approaching national systems. In *National systems of innovation: Towards a theory of innovation and interactive learning*, ed. B. Lundvall. London: Pinter.

Antcliff, V. 2005. Broadcasting in the 1990s: Competition, choice and inequality? *Media, Culture and Society* 27 (6): 841–59.

Archer, L., and M. Hutchings. 2000. "Bettering yourself"? Discourses of risk, cost and benefit in ethnically diverse, young working-class non-participants' constructions of higher education. *British Journal of Sociology of Education* 21 (4): 555–74.

Archibugi, D., and B. Lundvall. 2001. Introduction: Europe and the learning economy. In *The globalizing learning economy*, ed. D. Archibugi and B.-Å. Lundvall. Oxford: Oxford University Press.

Arora, A., and A. Gambardella. 1990. Complementarity and external linkages: The strategies of the large firms in biotechnology. *Journal of Industrial Economics* 38 (4): 361–79.

Ball, S. 2003. *Class strategies and the education market: The middle classes and social advantage*. London: Routledge Falmer.

Ball, S., and C. Vincent. 1998. "I heard it on the grapevine": "Hot" knowledge and school choice. *British Educational of Sociology of Education* 19 (3): 377–400.

Bearman, P. S. 1993. *Relations into rhetorics: Local elite structure in Norfolk, England 1540–1640*. New Brunswick, NJ: Rutgers University Press.

Beck, U. 1992. *Risk society: Towards a new modernity*. London: Sage.

Benner, C. 2003. Labour flexibility and regional development: The role of labour market intermediaries. *Regional Studies* 37 (6): 621–33.

Bergh, A., and G. Fink. 2005. *Escaping from mass education: Why Harvard pays*. Department of Economics Working Paper No. 2. Lund University, Department of Economics.

Black, D. A., and J. A. Smith. 2006. Estimating the returns to college quality with multiple proxies for quality. *Journal of Labor Economics* 24 (3): 701–28.

Black, D. A., J. A. Smith, and K. Daniel. 2005. College quality and wages in the United States. *German Economic Review* 6 (3): 415–43.

Blackley, M. 2005. Virgin Blue discriminated against older hostesses "for young blondes." *The Scotsman*, October 11.

Blair, H. 2001. "You're only as good as your last job": The labour process and labour market in the British film industry. *Work, Employment and Society* 15 (1): 149–69.

Blanden, J., P. Gregg, and S. Machin. 2005. Social mobility in Britain: Low and falling. *CentrePiece* Spring: 18–20.

Blundell, R., L. Dearden, and B. Sianesi. 2005. Evaluating the effect of education on earnings: Models, methods and results from the National Child Development Survey. *Journal of the Royal Statistical Society: Series A* 168 (3): 473–512.

Bok, D. 2003. *Universities in the marketplace: The commercialization of higher education*. Princeton, NJ: Princeton University Press.

Bollier, D. 2002. *Silent theft: The private plunder of our common wealth*. New York: Routledge.

Bourdieu, P. 1997. The forms of capital. In *Education: culture, economy, society*, ed. A. H. Halsey, H. Lauder, P. Brown, and A. S. Wells. Oxford: Oxford University Press.

Bourdieu, P., and J. Passeron. 1977. *Reproduction: In education, society and culture*. California: Sage.

Bourdieu, P., and L. Wacquant. 1992. *An invitation to reflexive sociology*. Chicago: University of Chicago Press.

Bowles, S., and H. Gintis. 2002. The inheritance of inequality. *Journal of Economic Perspectives* 16 (3): 3–30.

Breen, R., ed. 2004. *Social mobility in Europe*. Oxford: Oxford University Press.

Breitman, R. 2007. *Young women earn more than men in big U.S. cities. International Business Times*, August 5. http://www.ibtimes.com/articles/20070805/men-women-salaries.htm.

Brennan, J., and T. Shah. 2003. *Access to what? Converting education opportunity into employment opportunity*. London: Centre for Higher Education Research.

Brewer, D., E. Eide, and R. Ehrenberg. 1999. Does it pay to attend an elite private college? Cross-cohort evidence on the effects of college type on earnings. *Journal of Human Resources* 34 (1): 104–23.

Brint, S. 2001. Professionals in the "knowledge economy": Rethinking the theory of post industrial society. *Current Sociology* 49 (4): 101–32.

Brown, P. 2000. The globalisation of positional competition. *Sociology* 34 (4): 633–53.

Brown, P., and A. Hesketh. 2004. *The mismanagement of talent: Employability and jobs in the knowledge economy*. Oxford: Oxford University Press.

Brown, P., and H. Lauder. 1996. Education, globalisation, and economic development. *Journal of Education Policy* 11: 1–24.

———. 2001. *Capitalism and social progress: The future of society in a global economy*. New York: Palgrave.

Brown, P., and R. Scase. 1994. *Higher education and corporate realities: Class, culture and the decline in graduate careers*. London: UCL Press.

Brown, G. 2003. *A modern agenda for prosperity and social reform*. February 3, 2003. http://80.69.6.120/newsroom_and_speeches/press/2003/press_12_03.cfm

Brown, R. 2003. New Labour and higher education: Dilemmas and paradoxes. *Higher Education Quarterly* 57 (3): 239–48.

Burt, R. 1997. The contingent value of social capital. *Administrative Science Quarterly* 42: 339–65.

Calder, P. 2003. The hoard of the Rings. *The New Zealand Herald*, November 29. Available http://www.nzherald.co.nz/section/3/story.cfm?c_id=3&objectidl=3536699

Campbell, G. 1995. Town Hollywood. *Listener* February 19–23.

Carlsson, B., S. Jacobsson, M. Holmén, and A. Rickne. 2002. Innovation systems: Analytical and methodological issues. *Research Policy* 31 (2): 233–45.

Casteilla, E., H. Hwang, E. Granovetter, and M. Granovetter. 2000. Social networks in Silicon Valley. In *The Silicon Valley edge*, ed. C. Lee, W. Miller, M. Hancock, and H. Rowen. Stanford, CA: Stanford University Press.

Castells, M. 1996. *The rise of the network society*. London: Blackwells.

———. 1998. *The end of the millennium*, 2nd ed. Oxford: Blackwell.

Chevalier, A. N., and G. Conlon. 2003. *Does it pay to attend a prestigious university?* London: Centre for the Economics of Education.

Clark, H. *Maximising spin-offs from The Lord of The Rings: Questions and answers*. Government of New Zealand, http://www.executive.govt.nz/MINISTER/clark/lor/qa.htm.

Cohen, S., and G. Fields. 1999. Social capital and capital gains in Silicon Valley. *California Management Review* 41 (2): 108–30.

Coleman, J. 1988. Social capital in the creation of human capital. *American Journal of Sociology* 94 (Suppl.): S95–S120.

Collins, R. 1979. *The credential society: An historical sociology of education and stratification.* New York: Academic Press.

Conlon, G., and A. Chevalier. 2002. *Financial returns to undergraduates,* London: Centre for the Economics of Education.

Cooke, P. 2002. Regional innovation systems: General findings and some new evidence from biotechnology clusters. *Journal of Technology Transfer* 27 (1): 133–45.

Corak, M. 2004. Generational income mobility in North America and Europe: An introduction. In *Generational income mobility in North America and Europe,* ed. M. Corak. Cambridge: Cambridge University Press.

Cortada, J. W. 1993. *The computer in the United States: from laboratory to market, 1930 to 1960.* Armonk, NY: M.E. Sharpe.

Cranford, C. 2005. Networks of exploitation: Immigrant labor and the restructuring of the Los Angeles janitorial industry. *Social Problems* 52 (3): 379–97.

Cullen, M. 2007. *Transforming tertiary education and the NZ economy* http://www.beehive.govt.nz/speech/transforming+tertiary+education+and+nz+economy Friday, 12 October 2007.

Cutler, A., B. Hindess, P. Hirst, and A. Hussain. 1977. *Marx's "Capital" and capitalism today.* London: Routledge & Kegan Paul.

Dale, S., and A. Krueger. 2002. Estimating the payoff to attending a more selective college: An application of selection on observables and unobservables. *Quarterly Journal of Economics* 117 (4): 1491–525.

Dean, M. 1995. Governing the unemployed self in an active society. *Economy and Society* 24 (4): 559–83.

———. 1999. *Governmentality: Power and rule in modern society.* London: Sage.

Dekker, D. 2007. MacDiarmid valued sharing. *Dominion Post,* February 15, 7.

de Bruin, A., and A. Dupuis. 2004. Flexibility in the complex world of non-standard work: The screen production industry in New Zealand. *New Zealand Journal of Employment Relations* 29 (3): 35–66.

Department for Business, Enterprise, and Regulatory Reform. 2002. *Evaluation of the teaching company scheme—final report to the small business service.* http://www.berr.gov.uk/about/economics-statistics/economics-directorate/page21981.html

Department for Education and Employment. 2000. *Connexions. The best start in life for every young person.* Nottinghamshire: Department for Education and Skills.

Department for Education and Skills. 2003. *The future of higher education,* London: Department for Education and Skills.

———. 2005. *14–19 Education and skills—white paper,* Nottinghamshire: Department for Education and Skills.

Department for Trade and Industry 2003. Innovation report:Competing in the global economy: the innovation challenge. London: Department of Trade and Industry.

Department for Trade and Industry 2004. *Department of Trade and Industry—Five year programme: Creating wealth from knowledge.* London: Department for Trade and Industry.

Department of Trade and Industry. 2007a. *Knowledge transfer networks*. http://www .berr.gov.uk/dius/innovation/technologystrategyboard/tsb/technologyprogramme/ KTN/page12567.html

———. 2007b. *Knowledge transfer partnerships: Annual report 2005/06*. London: Department of Trade and Industry.

Department of Statistics. 2005. *Background to the screen production industry*, Wellington: Department of Statistics.

Devine, F. 2004. *Class practices: How parents help their children get good jobs*. Cambridge: Cambridge University Press.

Dex, S., J. Willis, R. Patterson, and E. Sheppard. 2000. Freelance workers and contract uncertainty: The effects of contractual changes in the television industry. *Work, Employment and Society* 14: 283–305.

DiMaggio, P., and W. Powell. 1983. The iron cage revisited: Institutional isomorphism and collective rationality in organizational fields. *American Sociological Review* 48 (2): 147–60.

Doloreux, D. 2002. What we should know about regional systems of innovation. *Technology in Society* 24: 243–63.

du Gay, P., ed. 1997. *Production of culture/cultures of production*. London: Sage.

Egerton, M. 2002. Higher education and civic engagement. *British Journal of Sociology* 53 (4): 603–20.

Eliasson, K. 2006. *The role of ability in estimating the returns to college choice: New Swedish evidence*. Department of Economics, Umeå University, http://ideas.repec .org/p/hhs/umnees/0691.html.

Emirbayer, M. 1997. Manifesto for a relational sociology. *American Journal of Sociology* 103 (2): 281–317.

Emmison, M. 2003. Social class and cultural mobility: Reconfiguring the cultural omnivore thesis. *Journal of Sociology* 39 (3): 211–31.

Erickson, B. 1996. Culture, class and connections. *American Journal of Sociology* 102 (1): 217–51.

Erikson, R., and J. Goldthorpe. 1992. *The constant flux: A study of class mobility in industrial societies*. New York: Open University Press.

Fernandez, R., E. Castilla, and P. Moore. 2000. Social capital at work: Networks and employment at a phone center. *American Journal of Sociology* 105 (5): 1288–356.

Fevre, R. 2000. Socializing social capital: Identity, work, and economic development. In *Social capital: critical perspectives*, ed. S. Barton, J. Field, and T. Schuller. Oxford: Oxford University Press.

Florida, R. 2002. The economic geography of talent. *Annals of the Association of American Geographers* 92 (4): 743–55.

Freeman, C. 1997. *The economics of industrial innovation*. Cambridge, MA: MIT Press.

Galindo-Rueda, F., O. Marcenari-Gutierrez, and A. Vignoles. 2004. *The widening socio-economic gap in UK higher education*, London: Centre for the Economics of Education.

Garnsey, E., and H. Lawton Smith. 1998. Proximity and complexity in the emergence of high technology industry: The Oxbridge comparison. *Geoforum* 29 (4): 433–50.

Giddens, A. 1998. *The third way*. Cambridge: Polity.

Glasson, J. 2003. The recent and rapid rise of the Oxfordshire high-technology economy: Accidental or planned? Paper presented at the *Planning Research Conference*, April 8–10, Oxford.

Gorard, S. 2005. Where shall we widen it? Higher education and the age participation rates in Wales. *Higher Education Quarterly* 59 (1): 3–18.

Gorard, S., and G. Rees. 2002. *Creating a learning society?* Bristol: Policy Press.

Gough, J. 2004. The relevance of the limits to capital to contemporary spatial economics: For an anti-capitalist geography. *Antipode* 36 (3): 512–26.

Government of New Zealand. 1995. *Education Act 1989. In Reprinted Act: Education [with Amendments Incorporated]*. 1989.

Granovetter, M. 1974. *Getting a job: A study of careers and contacts*. Cambridge, MA: Harvard University Press.

———. 1995. *Getting a job: A study of careers and contacts*. 2nd ed. Chicago: Chicago University Press.

Grieco, M. 1987. *Keeping it in the family: Social networks and employment*. London: Tavistock.

———. 1996. *Workers' dilemmas: Recruitment, reliability and repeated exchange: An analysis of urban social networks and labour circulation*. London: Routledge.

Groysberg, B., and R. Abrahams. 2006. Lift outs: How to acquire a high-functioning team. *Harvard Business Review* 84 (12): 133–40.

Halsey, A. 1993. Trends in access and equity in higher education: Britain in international perspective. *Oxford Review of Education* 19: 124–40.

Hamermesh, D., and J. Biddle. 1994. Beauty and the labor market. *American Economic Review* 84 (5): 1174–94.

Harrison, B. 1994. The small firms myth. *California Management Review* 36 (3): 142–58.

Hayek, F. 1945. The use of knowledge in society. *American Economic Review* 35 (4): 519–30.

Higher Education Funding Council. 2007. *Aimhigher*. London: Higher Education Funding Council.

Hirsch, F. 1976. *Social limits to growth*. Cambridge, MA: Harvard University Press.

H.M. Government. 2005. *Skills: Getting on in business, getting on at work*. London: H.M. Government.

H.M. Treasury. 2003. Lambert review of business-university collaboration. London: H.M. Treasury.

———. 2004. *Science & innovation investment framework 2004—2014*. London: H.M. Treasury.

Howells, J. 2005. Innovation and regional economic development: A matter of perspective? *Research Policy* 34: 1220–34.

Howells, J., and M. Nedeva. 2003. The international dimension to industry-academic links. *International Journal of Technology Management* 25 (1/2): 5–17.

Hudson, J. 2006. Inequality and the knowledge economy: Running to stand still? *Social Policy and Society* 5, no. 2: 207–22.

Hughes, A. 2006. *University-industry linkages and UK science and innovation policy.* Working Paper No. 326. Cambridge: Centre for Business Research, University of Cambridge.

Hughes, D., and D. Pearce. 2003. Secondary school decile ratings and participation in tertiary education. *New Zealand Journal of Educational Studies* 38 (2): 193–206.

James, R. 2002. *Socioeconomic background and higher education participation: An analysis of school students' aspirations and expectations.* Canberra: Department of Education, Science and Training.

James, E., N. Alsalam, J. C. Conaty, and T. Duc-Le. 1989. College quality and future earnings: Where should you send your child to college? *American Economic Review* 79 (2): 247–52.

Jeffcutt, P., and A. Pratt. 2002. Managing creativity in the cultural industries. *Creativity and Innovation Management* 11 (4): 225–33.

Johnson, D. 2002. It's a small(er) world: The role of geography in biotechnology innovation. Working paper No. 2002–01. Wellesley, MA: Wellesley College Department of Economics.

Johnson, P., and R. Reed. 1996. Intergenerational mobility among rich and poor: Results from the National Child Development Survey. *Oxford Review of Economic Policy* 12 (1): 127–42.

Jones, C. 2002. Signaling expertise: How signals shape careers in creative industries. In *Career creativity: Explorations in the remaking of work*, ed. M. Peiperl, M. Arthur, R. Goffee, and N. Anand. Oxford: Oxford University Press.

Jones, D. 2005. It's official—everyone in New Zealand has been thanked: Peter Jackson as creative industries "ring-leader." Paper presented at the European Group for Organizational Studies, Berlin.

Jordan, S., and R. Strathdee. 2001. The "training gospel" and the commodification of skill: Some critical reflections on the politics of skill in Aotearoa/New Zealand. *Journal of Vocational Education and Training* 53 (1): 391–405.

Kanter, R. 1995. *World class: Thriving locally in the global economy.* New York: Simon and Schuster.

Keep, E., and K. Mayhew. 1999. The assessment: Knowledge, skills, and competitiveness. *Oxford Review of Economic Policy* 15 (1): 1–15.

Keller, W. 2002. Geographic localization of international technology diffusion. *American Economic Review* 92 (1): 120–42

Kelly, K. 1999. *New rules for the new economy: 10 radical strategies for a connected world.* London: Fourth Estate.

Kerr, C., J. Dunlop, F. Harbison, and C. Myers. 1973. *Industrialism and industrial man: The problems of labour and management in economic growth.* Harmondsworth: England: Penguin.

Kettley, N. 2007. The past, present and future of widening participation research. *Bristish Journal of Sociology of Education* 28 (3): 333–47.

Kilduff, M., and D. Krackhardt. 1994. Bringing the individual back in: A structural analysis of the internal market for reputation in organizations. *Academy of Management Journal* 37 (1): 87–108.

Krugman, P., ed. 1986. *Strategic trade policy and the new international economics.* Cambridge, MA: MIT Press.

Lachnit, C. 2001. Employee referral saves time, saves money, delivers quality. *Workforce* June: 67–72.

Lam, A. 2002. Skills and careers of R & D knowledge workers in the network economy. Paper read at the *IIRA 13th World Congress*, Berlin, Germany.

Lamont, M., and V. Molnar. 2002. The study of boundaries in the social sciences. *Annual Review of Sociology* 28: 167–95.

Lauder, H. 2007. Towards a high skills economy: Higher education and the new realities of global capitalism. Presentation to the Victoria Management School and the Developing Human Capability Project.

Lauder, H., and Y. Mehralizadeh. 2001. Globalisation and the labour market. In *High skills: Globalisation, competitiveness and skill formation*, ed. P. Brown, A. Green, and H. Lauder. Oxford: Open University Press.

Leathwood, C. 2004. A critique of institutional inequalities in higher education. *Theory and Research in Education* 2 (1): 31–48.

Lee, C., W. F. Miller, M. Hancock, and H. Rowen. 2000. *The Silicon Valley edge: A habitat for innovation and entrepreneurship.* Stanford, CA: Stanford Business Books.

L.E.K. Consulting. 2006. *New Zealand biotechnology industry growth report.* Auckland: L.E.K Consulting.

Levine, D. 1999. *Choosing the right parents: Changes in the intergenerational transmission of inequality between the 1970s and early 1990s.* Berkeley: University of California Press.

Lin, N. 1999. Building a network theory of social capital. *Connections* 22 (1): 28–51.

———. 2001. *Social capital: A theory of social structure and action.* Cambridge: Cambridge University Press.

Lipsett, A. 2007. Firms turn to psychometric tests to pick graduate recruits. *Guardian Unlimited*, July 10.

Livingstone, D. 1999. Lifelong learning and underemployment in the knowledge society: A North American perspective. *Comparative Education* 35 (2): 163–86.

Lloyd, C., and J. Payne. 2002. The road to (no)where? The politics of "high skills" in the UK. Paper presented at the European Conference for Educational Research, Lisbon, Portugal.

Loogma, K., M. Umarik, and R. Vilu. 2004. Identification-flexibility dilemma of IT specialists. *Career Development International* 9 (3): 323–48.

Lundvall, B., ed. 1992. *National systems of innovation: Towards a theory of innovation and interactive learning.* London: Pinter.

Maharey, S. 2002. Education and work partnerships for the knowledge economy. Wellington: Parliamentary Speech Archive.

Maharey, S. 2004. Forward. In *Performance-based research fund—Evaluating research excellence: the 2003 assessment*. Wellington: Tertiary Education Commission.

Mallard, T. 2006. Beyond the valley of death: Making innovation work. Paper presented at Capitalising on Research, Auckland, New Zealand. http://www.beehive .govt.nz/node/27573

Marquand, D. 1997. *The new reckoning: Capitalism, states and citizens*. Cambridge: Polity Press.

Marsh, D. 2004. *Biotechnology in New Zealand description and analysis based on the 1998/99 and 2002 biotech surveys and a review of secondary sources*. Working paper in economics 1/04. Waikato: University of Waikato.

———. 2006. Evidence-based policy: Framework, results and analysis from the New Zealand biotechnology sector. *International Journal of Biotechnology* 8 (3/4):206–24.

Marshall, A. 1961. *Principles of economics*, 9th ed. London: Macmillan.

Martin, R. 1999. The new "geographical turn" in economics: Some critical reflections. *Cambridge Journal of Economics* 23: 65–91.

Massey, D. B., P. Quintas, and D. Wield. 1992. *High-tech fantasies: Science parks in society, science, and space*. London: Routledge.

Mayhew, K., C. Deer, and M. Dua. 2004. The move to mass higher education in the UK: Many answers to some questions. *Oxford Review of Education* 30 (1): 65–82.

McDonald, F., and G. Vertova, eds. 2002. *Cluster, industrial districts and competitiveness*. Aldershot: Ashgate.

McRobbie, A. 2002. From Holloway to Hollywood: Happiness at work in the new cultural economy. In *Cultural economy*, ed. P. du Gay and M. Pryke. London: Sage.

Minister of the Rings. 2001. *New Zealand Herald*, September 7, 6.

Ministry of Research Science and Technology. 2002. *New Zealand biotechnology strategy*. http://www.morst.govt.nz/publications/a-z/n/nz-biotechnology-strategy/.

Mizen, P. 1995. *The state, young people and youth training: In and against the training state*. London: Mansell.

Morley, L. 2007. The X factor: Employability, elitism and equity in graduate recruitment. *Twenty-First Century Society Journal of the Academy of Social Sciences* 2 (2): 191–207.

Morley, L., and S. Aynsley. 2007. Employers, quality and standards in higher education: Shared values and vocabularies or elitism and inequalities? *Higher Education Quarterly* 61 (3): 229–249.

Mulrooney, P. 2006. Creating his own special effect. *Dominion Post* July 20, 7.

Murphy, R. 1984. The structure of closure: A critique and development of the theories of Weber, Collins, and Parkin. *British Journal of Sociology* 35: 547–67.

———. 1988. *Social closure: The theory of monopolization and exclusion*. Oxford: Clarendon Press.

Nahapiet, J. and Ghoshal, S. 1998. Social capital, intellectual capital, and the organization of advantage. *The Academy of management Review* 23 (2): 242–65.

Nahapiet, J., and S. Ghoshal. 2000. Social capital, intellectual capital, and the organization of advantage. In *Knowledge and social capital*, ed. E. Lesser. Boston: Butterworth Heinemann.

Naidoo, R. 2003. Repositioning higher education as a global commodity: Opportunities and challenges for future sociology of education work. *British Journal of Sociology of Education* 24 (2): 249–59.

Naidoo, R., and I. Jamieson. 2005. Empowering participants or corroding learning? Towards a research agenda on the impact of student consumerism in higher education. *Journal of Education Policy* 20 (3): 267–81.

Naylor, R., J. Smith, and A. McKnight. 2002. *Sheer class? The extent and sources of variation in the UK graduate earnings premium.* London: Centre for Analysis of Social Exclusion, London School of Economics.

New Zealand Qualifications Authority 1996. *The national qualifications framework —issues.* Wellington: New Zealand Qualifications Authority.

New Zealand Ministry of Education. 2006. *Trends in the contribution of tertiary education to the accumulation of educational capital in New Zealand 1981–2001,* Wellington: New Zealand Ministry of Education.

New Zealand Screen Council. 2005. *Overview of the New Zealand screen production sector,* Wellington: New Zealand Screen Council.

Nickson, D., Warhurst, C. and Cullen, A. 2003. Bringing in the excluded: Aesthetic labour, skills and training in the 'new' economy. *Journal of Education and Work* 16 (2): 185–203.

Nie, N., J. Junn, and K. Stehlik-Barry. 1996. *Education and democratic citizenship in America.* Chicago: University of Chicago Press.

Nonaka, I., and N. Konno. 1998. The concept of "Ba": Building a foundation for knowledge creation. *California Management Review* 40 (3): 40–54.

North, D. 1990. *Institutions, institutional change, and economic performance.* Cambridge: Cambridge University Press.

Office of the Prime Minister. 2002. *Growing an innovative New Zealand.* Wellington: Office of the Prime Minister.

Organisation for Economic Cooperation and Development. 1996. *The knowledge-based economy.* Paris: Organisation for Economic Cooperation and Development.

Pavitt, K. 1984. Sectoral patterns of technical change: Towards a taxonomy and a theory. *Research Policy* 13 (6): 343–73.

Peters, M., and T. May. 2004. Universities, regional policy and the knowledge economy. *Policy Futures in Education* 2 (2): 263–77.

Potts, G. 2002. Regional policy and the "regionalization" of university-industry links: A view from the English regions. *European Planning Studies* 10 (8):987–1012.

Power, S., and G. Whitty. 2006. Education and the middle class: A complex but critical case for the sociology of education. In *Education, Globalisation and Social Change,* ed. H. Lauder, P. Brown, J.-A. Dillabough, and A. H. Halsey. Oxford: Oxford University Press.

Pring, R. 2005. Labour government policy 14–19. *Oxford Review of Education* 31 (1):71–85.

Putnam, R. 2000. *Bowling alone: The collapse and revival of American community*. New York: Simon and Schuster.

Qualifications and Curriculum Authority 2004. *Principles for a credit framework for England*. London: Qualifications and Curriculum Authority.

Qualifications and Curriculum Authority. 2005. *Framework for achievement: Questions and answers*. London: Qualifications and Curriculum Authority.

Reay, D. 2000. A useful extension of Bourdieu's conceptual framework?: Emotional capital as a way of understanding mothers' involvement in their children's education? *Sociological Review* 48 (4): 568–85.

Report of the Workplace Productivity Working Group. 2004. *The Workplace Productivity Challenge*, Wellington: Department of Labour.

Rey, S. J. 2001. Spatial empirics for economic growth and convergence. *Geographical Analysis* 33 (3): 206–14.

Rickards, T., and S. Moger. 2000. Creative leadership processes in project team development: An alternative to Tuckman's stage model. *British Journal of Management* 11 (4): 273–83.

Roizen, J., and M. Jepson. 1985. *Degrees for jobs: Employer expectations of higher education*. Guilford: Society for Research into Higher Education & NFER-Nelson.

Rosenbaum, J. 2002. *Beyond college for all: Career paths for the forgotten half*. New York: Sage.

Rumelt, R. P., D. Schendel, and D. J. Teece. 1991. Strategic management and economics. *Strategic Management Journal* 12 (Winter): 5–25

Sako, M. 1999. From individual skills to organizational capability in Japan. *Oxford Review of Economic Policy* 15: 114–26.

Saxenian, A. 2002. Transnational communities and the evolution of global production networks: The cases of Taiwan, China and India. *Industry and Innovation* 9 (3): 183–202.

Schumpeter, J. A. 1934. *The theory of economic development*. Cambridge, MA: Harvard University Press.

Scott, J. 1996. *Stratification and power: Structure of class, status and command*. Cambridge: Cambridge: Polity Press.

Scott, W.R. 1995. *Institutions and organizations*. Thousand Oaks, California: Thousand Oaks, California: Sage.

Shenkar, O., and E. Yuchtman-Yaar. 1997. Reputation, image, prestige, and goodwill: An interdisciplinary approach to organizational standing. *Human Relations* 50 (11): 1361–81.

Shenkar, O. 2004. *The Chinese century: The rising Chinese economy and its impact on the global economy, the balance of power, and your job*. Philadelphia: Philadelphia: Wharton School.

Simmie, J. 2003. Innovations for urban regions as national and international nodes for the transfer and sharing of knowledge. *Regional Studies* 37, no. 6 & 7: 607–620.

Simmie, J. and Sennett, J. 1999. Innovative clusters: global or local linkages. *National Institute Review* 170, no. 87–102.

Smetherham, C. 2006. First among equals? Evidence on the contemporary relationship between educational credentials and the occupational structure. *Journal of Education and Work* 19, no. 1: 29–45.

Smith, J., McKnight, A. and Naylor, R. 2000. Graduate employability: Policy and performance in higher education in the UK. *Economic Journal* 110, no. 464: F382–411.

Snow, C. 1964. *The two cultures and a second look.* Cambridge: Cambridge: Cambridge University Press.

Solon, G. 1992. Intergenerational income mobility in the United States. *American Economic Review* 82, no. 3: 393–408.

Strathdee, R. 2003. The qualifications framework in New Zealand: Reproducing existing inequalities or disrupting the positional conflict for credentials. *Journal of Education and Work* 16 (2): 147–64.

———. 2005a. Globalisation, innovation and the declining significance of qualifications led social and economic change. *Journal of Education Policy* 20 (4): 437–56.

———. 2005b. *Social exclusion and the remaking of social networks.* Aldershot: Ashgate.

———. 2006. The creation of contrasting education and training markets in England and New Zealand. *Journal of Education and Work*, no.

———. 2007. School improvement, pre-service teacher education and the construction of social networks in New Zealand and England. *Journal of Education for Teaching* 33 (1): 19–33.

Strathdee, R., and D. Hughes. 2006. Socio-economic status and tertiary attendance in New Zealand. *New Zealand Journal of Educational Studies.* 41(2): 293–305.

Szreter, S. 1998. *A new political economy for New Labour: The importance of social capital.* PERC Policy Paper 15. Sheffield: Political Economy Research Centre.

Tepper, S. 2002. Creative assets and the changing economy. *Journal of Arts Management, Law, and Society* 32 (2): 159–68.

Tertiary Education Commission. *Partnerships for excellence: Successful applicants from the 2004–06 funding round.* 2004. http://www.tec.govt.nz/funding/strategic/p4excellence/successful-applicants.htm.

———. 2007 *Tertiary Education Strategy 2007–12: incorporating Statement of Tertiary Education Priorities 2008-10.* Wellington: Tertiary Education Commission.

Thrupp, M. 2001. School-level education policy under New Labour and New Zealand Labour: A comparative update. *British Educational of Educational Studies* 49 (2): 187–212.

Tilly, C. 1998. *Durable inequality.* Berkeley: University of California Press.

Tizard, J. *Cultural sector in great shape one year on from Cultural Recovery Package.* May 18, 2001. http://www.executive.govt.nz/Speech.aspx?type=press&rid=34681.

Waring, M., A. Hodgson, C. Savory, and K. Spours. 2003. *Curriculum 2000 and higher education: Villains or victims?* Broadening the Advanced Level Curriculum: IOE/Nuffield Series Number 8.

Weeden, K. and Grusky, D. 2005. Are There Any Big Classes at All? Research in Social Stratification and Mobility 22: 3–56.

Whelan, C., and R. Layte. 2002. Late industrialisation and the increased merit selection hypothesis. *European Sociological Review* 18 (1):35–50.

Wolf, A. 2002. *Does education matter: Myths about education and economic growth.* London: Penguin.

Wong, S., and J. Salaff. 1998. Network capital: Emigration from Hong Kong. *British Journal of Sociology* 49 (3): 358–74.

Woolcock, M., and D. Narayan. 2000. Social capital: Implications for development theory, research and policy. *World Bank Research Observer* 15 (2): 225–49.

Woolf, M., and S. Holly. 1994. *Employment patterns and training needs 1993/94*, London: Skillset.

Wu, C. 1998. Embracing the enterprise culture: Art institutions since the 1980s. *New Left Review* 230: 28–57.

Wynn, M. and Jones, P. 2006. Delivering business benefits through Knowledge Transfer Partnerships. *International Journal of Entrepreneurship and Small Business* 3, no. 3/4: 310–20.

Wynn, M., Jones, P., Roberts, C. and Little, E. 2008. Innovation in the construction and property management industries: Case studies of the knowledge transfer partnership scheme. *Property Management* 26, no. 1: 66–78.

Zhang, L. 2005. Do measures of college quality matter? The effect of college quality on graduates' earnings. *Review of Higher Education* 28 (4): 571–96.

Zucker, L. 1986. Production of trust: Institutional sources of economic structure, 1840–1920. *Research in Organisational Behaviour* 8: 53–111.

Zucker, L., and M. Darby. 1996. Star scientists and institutional transformation: Patterns of invention and innovation in the formation of the biotechnology industry *Proceedings of the National Academy of Sciences of the United States of America* 93: 12709–16.

Zucker, L., M. Darby, and M. Brewer. 1998. Intellectual human capital and the birth of U.S. biotechnology enterprises. *American Economic Review* 88 (1):290–306.

Index